PENGUIN BOOKS

MOA

Richard Wolfe has worked as display artist at Canterbury Museum and Auckland War Memorial Museum, and is now a freelance curator and writer. His books include several on kiwiana (co-authored with Stephen Barnett), along with *Well Made New Zealand – A Century of Trademarks*, *The Way We Wore*, *All Our Own Work – New Zealand's Folk Art* and *Kiwi: More than a Bird*. His five children's books, all illustrated by his wife, artist Pamela Wolfe, include *Midnight at the Museum*. He lives in Freemans Bay, Auckland.

# MOA

## The dramatic story of the discovery of a giant bird

Richard Wolfe

PENGUIN BOOKS

PENGUIN BOOKS

Penguin Books (NZ) Ltd, cnr Airborne and Rosedale Roads, Albany,
Auckland 1310, New Zealand
Penguin Books Ltd, 80 Strand, London, WC2R 0RL, England
Penguin Group (USA) Inc., 375 Hudson Street, New York, NY 10014, United States
Penguin Books Australia Ltd, 250 Camberwell Road, Camberwell,
Victoria 3124, Australia
Penguin Books Canada Ltd, 10 Alcorn Avenue, Toronto,
Ontario, Canada M4V 3B2
Penguin Books (South Africa) (Pty) Ltd, 24 Sturdee Avenue, Rosebank,
Johannesburg 2196, South Africa
Penguin Books India (P) Ltd, 11, Community Centre, Panchsheel Park,
New Delhi 110 017, India

Penguin Books Ltd, Registered Offices: 80 Strand, London, WC2R 0RL, England

First published by Penguin Books (NZ) Ltd, 2003

1 3 5 7 9 10 8 6 4 2

Designed by Mary Egan
Typeset by Egan-Reid Ltd, Auckland
Printed in Australia by McPherson's Printing Group

Front cover images: Lesser megalapteryx, chromolithograph after an oil painting by
George Edward Lodge, reproduced by permission of the Topham Picture Library. Moa
bone, lithograph by G. Sharf from *Transactions of the Zoological Society of London*
1842, reproduced by kind permission of the President and Council of the Royal College
of Surgeons.

Back cover images: Right foot of a moa, hand-coloured lithograph by J. Dinkel from
Gideon Mantell's *Pictorial Atlas of Fossil Remains*, 1850. Moa skull, lithograph by Joseph
Smit from *Transactions of the Zoological Society of London*, 1870, reproduced
by kind permission of the President and Council of the Royal College of Surgeons.
Dr Richard Owen standing beside moa skeleton, photograph by F. E. McGregor,
reproduced by kind permission of the Canterbury Museum.

ISBN 0 14 301873 6
A catalogue record for this book is available
from the National Library of New Zealand.

www.penguin.co.nz

# CONTENTS

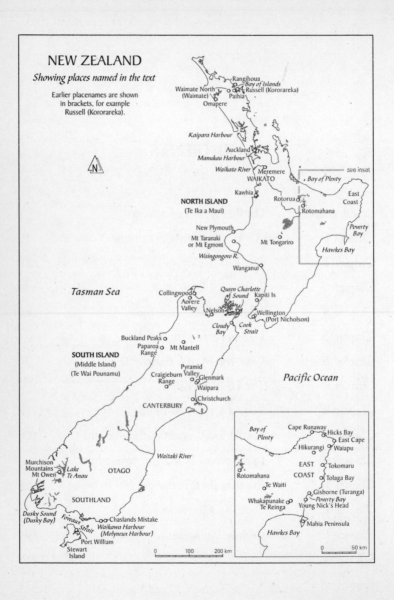

# NEW ZEALAND

*Showing places named in the text*

Earlier placenames are shown
in brackets, for example
Russell (Kororareka).

Rangihoua
Bay of Islands
Waimate North · Russell (Kororareka)
(Waimate) Paihia
Omapere

Kaipara Harbour

Auckland
Manukau Harbour

Waikato River Meremere
WAIKATO

*see inset*

Bay of Plenty

East
Coast

Kawhia
Rotorua

**NORTH ISLAND**
(Te Ika a Maui)

Rotomahana

Poverty
Bay

New Plymouth

Mt Taranaki
or Mt Egmont

Mt Tongariro

Hawkes Bay

Waingongoro R.

Wanganui

*Tasman Sea*

Collingwood

Queen Charlotte
Sound Kapiti Is

Aorere
Valley
Nelson

Wellington
(Port Nicholson)

Cloudy Cook
Bay Strait

Buckland Peaks

Paparoa Mt Mantell
Range

**SOUTH ISLAND**
(Middle Island)
(Te Wai Pounamu)

Pyramid
Valley
Craigieburn Glenmark
Range
Waipara
Christchurch

*Pacific Ocean*

CANTERBURY

Waitaki River

Murchison
Mountains Lake
Mt Owen Te Anau

OTAGO

SOUTHLAND

Dusky Sound
(Dusky Bay)
Chaslands Mistake
Waikawa Harbour
(Molyneux Harbour)
Port William
Stewart
Island

0  100  200 km

## Inset

Bay of
Plenty
Cape Runaway Hicks Bay
East Cape
Hikurangi Waiapu

EAST Tokomaru
Rotomahana COAST Tolaga Bay

Te Waiti

Whakapunake Gisborne (Turanga)
Te Reinga Poverty Bay
Young Nick's Head

Mahia Peninsula

Hawkes Bay

0  50 km

# INTRODUCTION
## On the Wings of a Moa

*I had also many specimens of . . . bones which we all regarded as the rudimentary wings of the moa . . . all now vanished in the flames . . .*
SIR GEORGE GREY[1]

I encountered my first moa at an early age in one of the provincial museums of New Zealand, and clearly recall feeling dwarfed by that dusty wired-together skeleton in a glass case. In 1973 I joined the display staff of the Canterbury Museum, where a century earlier founding director Sir Julius von Haast had been the 'father' of moa research in New Zealand.[2] I learned about Glenmark Swamp, Pyramid Valley and other graveyards which had enriched that institution's huge collection of moa remains. A few years later, on the staff at Auckland Museum, I marvelled at a fully-feathered (thanks to emu) moa reconstruction, a top visitor attraction there since 1913. I also became interested in the moa's small surviving relative which had both given its name to New Zealanders and inspired their popular culture, 'kiwiana'. Whilst researching for a book on the kiwi as a national symbol I became aware of a photograph of craggy English anatomist Richard Owen holding a small piece of bone. It was on the basis of that slender evidence that in 1839 he boldly announced the existence of the moa to members of the Zoological Society of London.

Wondering whether the story of that modest fragment might be a suitable basis for a book, in September 2001 I determined to inspect it at its home since 1873, the Palaeontology Department of The Natural History Museum, London. I was not disappointed. That historic specimen looked every bit as 'unpromising' as members of the Zoological Society had described it when they first saw it 162 years earlier. It seemed remarkable that this humble relic had managed the journey from New Zealand to London in the first place, quite apart from being the first known scientific evidence of the tallest bird – at least when its neck was fully extended in the upright position – that ever walked the earth.

The famous bone brought out both the best and the worst in people. Richard Owen's original statement came to be regarded by colleagues as nothing short of a brilliant piece of deduction, although there remains the possibility that he knew more than he let on. Soon afterwards, the moa became the subject of bitter rivalry, as various claims and counter-claims were made by those seeking credit for its discovery. That this story occurred at all is due also to both the initiative of trader John Williams Harris, and the persistence shown by surgeon John Rule in the face of a doubting Professor Owen. Of course, had Joel Polack taken to London in 1838 some of the bones he claimed he saw during his earlier adventures on New Zealand's East Coast, the story of the discovery of the moa might have turned out very different.

After being the subject of a huge and unknown number of meals, the moa suffered the further indignity of being largely forgotten. Later, when it was discovered by Europe, its remains got a second going over. The extinct bird then tested the limits of science, and at the same time provoked some unusual behaviour amongst its followers and alleged discoverers. In the meantime, the bone that started it all was unable to attract the interest of a museum, and so disappeared into a private collection for the next three decades. By the time it was rediscovered, the bird itself was on its way to becoming a standard feature in natural history museums around the world.

In researching and writing this biography of a bone I am grateful to the large number of individuals and institutions who provided assistance and advice. In particular I thank Geoff Walker at Penguin Books (NZ) and other Penguin staff who have given form to my manuscript. I would also like to thank John Buckland; Brian Gill, ornithologist, Auckland War Memorial Museum; Graham Turbott, Director Emeritus, Auckland War Memorial Museum; Roger Neich, ethnologist, Auckland War Memorial Museum; Bryan Harold and staff at Moa-Hunter Books, Ponsonby, Auckland; Russell Jackson; Sandra Chapman, Palaeontology Department, The Natural History Museum, London, for providing access to the bone; Michael Palmer, archivist, Zoological Society of London; Frank Rickarby, The British Library, London; J. Pitcher and F. S. McHenry, Newspaper Library, The British Library, London; staff at the Royal Geological Society of London; Simon Chaplin, Tina Craig and staff at the Hunterian Museum and Library, Royal College of Surgeons, London; Staff at the Print Room, Victoria and Albert Museum, London; Andrew Potter, Library, Royal Academy, London; Parliamentary Archives, House of Lords Records Office, London; Christopher Johnstone; Glenda Daff, Frankston Library Service, Frankston, Victoria, Australia; David Flegg, State Library of Victoria, Australia; Adrienne Simpson, H. B. Williams Memorial Library, Gisborne; Max Cryer; Staff at the Auckland Research Centre and Special Collections, Auckland Central Library; Averil E. Callisen, Research Centre, Alexander Turnbull Library, Wellington; Ponsonby Meats, Auckland; David Spalding; Susan Snell, Archives, The Natural History Museum, London; Mere Roberts, School of

Geography and Environmental Sciences, University of Auckland; Iain Buchanan, Art History Department, University of Auckland; Carol Gokce and Emma Bennett, Library, The Natural History Museum, London; Jenny Newell, Ethnology Department, The British Museum, London; Paul Brunton and Marie Alcorn, State Library of New South Wales, Sydney; Helen Bersten, archivist, Sydney Jewish Museum, Sydney; Eddie Sun, Selwyn Library, St John's College, Auckland; Emily, Stuart and Gwen for the accommodation in Hoxton, and my wife Pam for her constant patience and support during this latter-day moa hunt.

Because this book is primarily concerned with the story of the first identified moa bone, it does not deal with the more complex and scientific issues relating to that bird. The number of species, its posture and date of extinction, for example, have been dealt with in such authoritative volumes as *Prodigious Birds* by Atholl Anderson and, more recently, *The Lost World of the Moa* by Trevor Worthy and Richard Holdaway, and will doubtless continue to be subjects of much debate. In fact, the Otago Museum recently announced it was preparing to 'launch an expedition to a remote mountain area where hunters found ancient moa bones last year'.[3] Apart from the fact that this nine-strong scientific party was planning to fly in by helicopter, that report could just have easily been from 1853 as 2003. The Museum already had nine of the 11 known moa species, and hoped this expedition might produce bones from a tenth. But those numbers might be about to change, for the same newspaper reported that Dunedin-born Oxford University zoologist Alan Cooper was working on moa DNA samples, and his results might either increase or reduce the number of species.

The one certain thing about the moa is that it has gone, and while recent generations of New Zealanders may feel absolved of any blame, this loss provides an increasingly important lesson for us all. Even in death the moa contributed much to this country, especially in the areas of scientific development and national identity. We therefore owe it to that big bird that its surviving relatives and other members of our unique fauna do not go the same way.

Richard Wolfe
February 2003

# 1

## AN UNPROMISING FRAGMENT
### A Small Matter of Identification

*The texture of the bone, which affords the chief evidence of its ornithic character, presents an extremely dense crust, varying from one to two lines in thickness; then there occurs a lamello-cellular structure of two to three lines in thickness.*

RICHARD OWEN, 1840[1]

By the middle of the 18th century London had overtaken Paris and Constantinople to become the largest city in the world. Before long, stimulated by developments that included the imposition of a gin tax, the building of hospitals and the introduction of vaccinations, the city's population began a dramatic increase. By 1801 it had passed the first million mark, and after another forty years of exponential growth had doubled to reach the second. London was now the capital of an expanding empire and centre of international trade and finance, its teeming streets filled with horse-drawn traffic and citizens who could still enjoy the spectacle of dogfights, cockfights and public hangings. Wooden paving had recently been laid on some streets in an attempt to muffle the increasing clatter of the first Victorian metropolis.[2]

On the afternoon of 18 October 1839 Dr John Rule set out from No. 66 Fetter Lane, heading west across central London's Holborn and Chancery Lane. Two centuries earlier, Fetter Lane had been the haunt of one Isaac Praisegod Barebone, a leather-seller whose middle name proudly proclaimed his primary calling as a fiery preacher. This god-botherer was also something of a local nuisance, and his skeletal surname now hinted at Dr Rule's reason for passing through that same neighbourhood. Fetter Lane had always been something of a boundary, defining parishes and marking both the edge of the old City of London and the point where the Great Fire had stopped.[3] Rule now carried with him a small parcel, one whose contents would shortly alter another sort of boundary, within the altogether different world of natural history.

John Rule was on his way to Lincoln's Inn Fields, London's largest square and covering 12 acres. On its southern side stood an eponymous tribute to the achievements and accumulations of one man – John Hunter. He had arrived in London from Scotland in 1748 as a medical student, and went on to develop the art of dissection and become the greatest scientific surgeon of his day. During his lifetime, surgery progressed from a mysterious activity carried out by barbers to a scientific profession, but it was hardly one for the faint-hearted, for amputations then took place without the benefit of anaesthesia. The new science of medicine was necessarily advanced by means of post-mortem operations, as was the understanding of the animal kingdom.

Whilst serving as an army surgeon in Portugal during the Seven Years' War between Britain, France and Spain, Hunter honed his wide-ranging scientific interests. Among other things these encompassed the local geology, the effects of gunshot wounds, the sharpness of hearing in fish and the regeneration of lizards' tails.[4] On his return to England in 1763 these studies were the basis of his burgeoning collection of specimens, a graphic illustration of his theory that living things were demonstrably capable of constant adaptation. In 1776 Hunter was appointed surgeon to the King, but he was also famous for his temper, and in 1793 he died after suffering a fit brought on by an argument. England was now at war again, and there was little interest in securing Hunter's massive collection of specimens for the nation. The prospect of putting up the money reportedly caused Prime Minister Sir William Pitt to complain that Treasury hardly had enough to buy bullets, let alone bottles of pickled specimens.[5] But six years later, after strenuous efforts by his executors and former pupils, Parliament found the funds and Hunter's medical legacy was handed over to the custody of what shortly became known as the Royal College of Surgeons.

The Hunterian Museum opened in 1813, initially only on Tuesdays and Thursdays in May and June, and restricted to 120 visitors per day. By 1834 it needed enlarging and so the College engaged architect Sir Charles Barry, who would shortly also design new Houses of Parliament to replace those destroyed by fire that same year.[6] The new Hunterian, twice the previous size, opened in 1837, with three floors devoted to the sciences of physiology and pathology. The ground floor was ringed by fluted columns and occupied by rows of glass-topped display cabinets and larger specimens on plinths. The latter included representatives of man and beast – in particular the skeleton of the 'Irish Giant', Charles Byrne, (also referred to as O'Brian) who reputedly reached a height of 8 feet 4 inches in life. By way of contrast, he shared his podium with the osseous remains of Caroline Crachami, the 'Sicilian Dwarf', who stood barely 20 inches high and needed the protection of a glass dome.[7]

Overlooking proceedings from the end of the hall, and looming above a bust of founder John Hunter, was the skeleton of Chunee, a five-ton Indian elephant. Once the star of a local menagerie, this animal had threatened to break out from its cage in a fit of anger whilst being led along the Strand in 1826. This violent temper tantrum was put down to 'annual paroxysm' or sexual musth, the period of aggressiveness in the males of the species, which in Chunee's case had been aggravated by inflammation of one of his tusks. A firing squad was hailed from nearby Somerset House to quell the enraged beast but their barrage proved ineffective, as did a small cannon fired from close range. In the end it fell to the keeper to dispatch his charge with a spear. But death made little difference, for the perforated pachyderm was soon back on public display. Shortly, his condition demanded he be sold off as meat – 11,000 pounds of it – and it was presumably at this point that the rapidly ripening carcass was found to contain over 100 musket balls. In 1831 the articulated skeleton of what was then the largest elephant ever exhibited in England was purchased by the Hunterian Museum at a cost 'commensurate with its rarity and perfection'.[8] In keeping with the instructional nature of the institution, the root cause of Chunee's discomfort, the base of the inflamed tusk showing a spicule of ivory projecting into the pulp, went on display alongside. In life Chunee had withstood a fusillade of bullets and cannon-balls, but in death he eventually succumbed to an even more determined attack. In 1941 his skeleton was destroyed by fire when a German bomb was dropped on the Hunterian.

Other exhibits testified to extremes of natural development and personal experience. Falling into the latter category was the thorax of one Thomas Tipple, who in 1812 had been impaled by the shaft of a chaise, the result of an unruly horse. Fortunately, two passing veterinary surgeons extricated Tipple from his predicament by withdrawing the invasive shaft. He lived another 11 years and became the subject of a medical publication – suitably entitled 'Tipple's Case of Recovery'. Naturally, the shaft which had caused his 'extraordinary wound' was also on exhibition.[9]

John Rule had not come to view any such morbid attractions however. He was following up a letter he had written earlier that same day to the Assistant Conservator of the Hunterian, Professor

Richard Owen. If Rule's appearance at the Hunterian seems a rather hasty follow-up, he was simply taking advantage of London's six daily mail deliveries and assuming his letter had been delivered and read. His foregoing communication came straight to the point:[10]

> Sir,
> Herewith I desire to offer for Sale a portion or fragment of a Bone, I believe the largest and most rare that has been found; part of the femur of a Bird now considered to be wholly extinct. It is unchanged after a lapse of many years – it may be centuries of years in which it has been imbedded in the mud of a river that disembogues into one of the Bays in New Zealand.

Before providing further information on this bone, Rule gave his opinion on the trustworthiness of the original source of his information – albeit at least second-hand – the native (Maori) traditions of New Zealand. He considered these people 'the most careful of any' when it came to 'preserving their traditions free from the embellishments of the imagination'. Rule was convinced he had good reason to believe this bone was a relic of a large and extinct bird of flight, and he presumed that its like had been found 'in no other place on the globe'. Indeed, he had personally seen nothing like it in the 'excellent and extensive collections in the Museums in London'. Such investigations had further convinced him that this fragment had belonged to a bird – 'and not a beast, nor a fish, nor amphibious animal'. He also recorded the bone's vital statistics – a length of six inches, a smaller circumference of five and a half inches, and a weight of seven and three-quarter ounces – in addition to the price he had in mind: ten guineas. He also included another letter, from J. W. Harris and dated Sydney, 28 February 1837, which had come to him with the bone.[11] This was the sole source of Rule's information, and in it Harris explained that according to native tradition the bone had 'belonged to a Bird of the Eagle kind, but which has become extinct' and was, curiously, called 'A Movie'. Harris also briefly described four other items, all Maori artefacts, which he passed on to Rule at the same time. Rule gave Owen no indication that he was about to receive a visit from the writer, and offered no further information on his own identity, simply signing his letter 'Your obedient servant / John Rule'.

Richard Owen was then 35 years old, a tall and ungainly individual with a massive head, lofty forehead, and curiously round, prominent and expressive eyes – all topped by long, lank, dark hair.[12] He was also a busy man, for in addition to being Assistant Conservator he was Hunterian Professor of Comparative Anatomy and Physiology at the Royal College of Surgeons. And now, with the sudden appearance of John Rule, he was about to get even busier. The two men met, and Owen's unannounced visitor disclosed one of the contents of his parcel, a small piece of bone. This merely confirmed Owen's suspicions, and he not unreasonably dismissed the item as nothing more than a piece of marrow-bone 'like those brought to table in a napkin'.[13] But Rule was not about to give up so easily, and took it upon himself to point out certain features of his specimen, in particular the unusual texture of the interior. This was a bold move indeed, forcing a leading comparative anatomist to endure the indignity of having the finer points of his discipline explained to him by a complete stranger. As Hunterian Professor, Owen had other commitments and was about to deliver a public lecture when Rule turned up. It is not difficult to imagine the imperious Professor hovering impatiently in his academic gown, waiting for this troublesome visitor to take the hint and realise the inconsequential nature of his scruffy offering.

Rule apparently saw no need to advise Owen of the fact that he too was a qualified medical man and surgeon, having become a Member of the Royal College of Surgeons in 1806. He had retired from naval surgery duties in 1825, and eight years later accepted a post as medical charge on an emigrant ship to Australia. In 1839 he was back in London, and in possession of what he believed was a relic of a bird hitherto unknown to science. If his own medical experience had assisted with some rudimentary identification of the bone, he had also undertaken some other research, so it was a fairly confident Dr Rule who now confronted an extremely sceptical Professor Owen.

Museum curators know from experience that specimens brought to them by the curious public are frequently neither as old nor as significant as their owners would like to believe. Owen therefore had reasonable grounds for dismissing the unprepossessing object in front of him. Rule, on the other hand, would surely have recognised a marrow or beef bone when he saw one and, equally, would not have troubled a notable scientist with such a trifle. But if it was bovine, as Owen insisted, then it was unlikely to be from New Zealand, a country that prior to European settlement had known no large land-based mammals. Owen also pointed out that even if it was a bird bone, then it most certainly could not be from one of the flying 'eagle kind'. Another (extinct) New Zealand bird that Owen did not yet know about was the largest eagle that ever existed, but even this did not produce bones of such dimensions. The Professor ventured that there had simply been confusion as to the bone's origin, whereupon Rule revealed another of the items in his parcel, a greenstone mere, or Maori war club. This disclosure was the clincher, for Owen recognised it as a weapon 'peculiar to the New-Zealanders' and it suggested that Rule really did have a connection with that country. Owen now offered to examine the 'unpromising fragment' of bone at his leisure, suggesting that if Rule returned the next day he would be able to provide a more considered opinion.

His lecture over, Richard Owen was able to turn his mind to the matter of Rule's bone. He examined its unusually textured interior, observing that it was unlike that of the long bones of known mammals. He then compared it with other specimens in the Hunterian Museum, beginning with the skeleton of an ox and expecting to quickly 'verify [his] first surmise'[14] and settle the matter there and then. However, there were 'precluding differences', as there were with various other quadrupeds whose remains he thought might have been found in New Zealand. The thorough Owen then worked through all other likely specimens at his disposal, systematically considering – and eliminating – the horse, ass, camel, llama, buffalo, bear, lion, dog, kangaroo, orang-utan, emu, stilt-bird, tortoise and even the human. Eventually, he came to the conclusion that the bone was the shaft of a femur, or thigh bone, and from a bird similar in size to the Hunterian's adult male ostrich. But whereas the femur of that bird contained air, or was 'pneumatic', this bone had once been filled with marrow. Therefore, Owen's initial suggestion that it was a marrow-bone was correct, except that he was thinking in terms of it belonging to a mammal rather than a large bird.[15]

John Rule duly turned up next day and heard the good news. Owen explained how he had been led to the only possible conclusion, and with the fragment in hand he gave Rule a guided tour of the skeletons on display, explaining his extensive deductive process. With this happy outcome, Rule now anticipated that the Royal College of Surgeons would be a willing buyer for his bone. Owen certainly encouraged him in this expectation, promising to recommend that the Museum Committee purchase it for the ten guineas Rule had in mind. However, one member had already described the bone as 'an unpromising fragment'[16] and others were not entirely convinced of Owen's judgement, so the Committee – which, incidentally, included William Clift, Owen's own father-in-law – decided not to purchase. Eventually the bone did find a buyer – at a much reduced price – but in the meantime was being readied for its next step in the scientific process.

If the fragment had failed to impress the authorities at the

Royal College of Surgeons, Owen was undeterred, and now sought Rule's permission to present it to the world of science. By so doing he was conforming to a tradition that had its origins in 1660 with the establishment of the first learned society in Great Britain, the Royal Society of London for Improving Natural Knowledge – later abbreviated to the Royal Society. It was to be followed by a host of others, including the Linnean Society (founded 1788), the Geological Society (1807), the Royal Astronomical Society (1820), the Royal Statistical Society (1830) and the Zoological Society of London (1826). There were also the Entomological and Wernerian Societies, the Royal Irish Academy, the Botanical Society of Edinburgh, the Dublin Natural History Society, the Lunar Society of Birmingham and the more lengthily named Natural History Society for the West Riding of Yorkshire and the Tweedside Physical and Antiquarian Society.[17] The advancement of 'natural knowledge' – now increasingly referred to as science – was the aim of all. Many of their disciplines overlapped, as did their membership, which was also heavily weighted towards professional and retired persons. For them, science was mostly an amateur pursuit, and spared the daily grind of the labouring classes they could devote themselves to self-improvement amongst like-minded company. Richard Owen, then one of the few truly professional scientists, was known to several such bodies, and in late October 1839 he decided to take his most recent 'find' to the next meeting of the Zoological Society of London.

England's early scientific advancement had been well served with its Royal and Linnean Societies, but by the early 1800s they exhibited some obvious shortcomings. There was mounting feeling that the Linnean Society was not devoting the same sort of energy to the promotion of zoology as it was to botany. Meanwhile, the highly influential, if not tyrannical, Sir Joseph Banks presided over the Royal Society for 41 years, and apparently did little for the growth of science in general. More specialist organisations were needed to promote the national importance of science, and it was out of this dissatisfaction that other learned bodies were born. These included the Zoological Society of London, whose immediate objective was 'the collection of such living subjects of the Animal kingdom as may be introduced and domesticated with advantage in this country'.[18] Thanks largely to the efforts of Richard Owen, it would soon extend its interest to certain extinct species as well.

In the late 1830s the fortnightly meetings of the Zoological Society were held at 28 Leicester Square, premises which coincidentally had once been the home and museum of John Hunter. On the evening of 12 November 1839, vice president and founding member William Yarrell was in the chair. He had begun his working life as a banking clerk, and a course of instruction in anatomy had equipped him to write several papers and a book on British birds which became the standard work on the subject. The evening got under way with matters of correspondence, beginning with a letter from Lady Shelley describing the antics of a black spider monkey that had recently been presented to the Society. Members were then advised of the imminent arrival of some live animals dispatched from Tunis: an antelope, three Numidian cranes and a young lynx. Another letter came from a member in Gibraltar who sent some specimens of fish, but it was perhaps the next item on the agenda that was the highlight of the evening. As dutifully recorded in the Society's *Proceedings*:

Professor Owen exhibited the bone of an unknown struthious bird of large size, presumed to be extinct, which had been placed in his hands for examination by Mr Rule, with the statement that it was found in New Zealand, where the natives have a tradition that it belonged to a bird of the Eagle kind, but which has become extinct, and to which they give the name 'Movie'.

21

The bone fragment, from what he now identified as a struthious bird – one of the *Struthioniformes* order, within the larger ratite group – had obviously been a focus of Owen's attentions over the previous three weeks. He now proceeded to outline its basic dimensions to the assembled members, before moving into more serious technical terminology. The outer surface, for example, was marked by 'very shallow reticulate indentations' and ridges – or *'lineae asperae'*. On the other hand, the interior was distinguished by a lamello-cellular structure, or cancellous texture, which closely resembled that of the femur of the ostrich. The elephant and crocodile had therefore been eliminated during the identification process because their bones were solid and non-cancellous respectively. By his own admission, Owen's interest in this relic of a 'large struthious bird in New Zealand' was stimulated by his knowledge that the same country was already home to 'one of the most extraordinary and anomalous' of such birds – a reference to what would soon be known as the kiwi. He saw an analogy between his new discovery and the extinct dodo of Mauritius, and suggested the two birds had similar proportions. Owen ended his paper on a suitably dramatic note, putting his professional credibility on the line: '[S]o far as my skill in interpreting an osseous fragment may be credited, I am willing to risk the reputation for it on the statement that there has existed, if there does not now exist, in New Zealand, a Struthious bird nearly, if not quite, equal in size to the ostrich.' The meeting then concluded with the exhibition of assorted skins of animals and birds from Tripoli, and specimens of the Portuguese man-of-war which had been netted off the Devonshire coast.[19]

The bone was now out of the bag, having been publicly presented, and the text of Owen's presentation of 12 November would shortly be described in a scientific journal, albeit the rather perfunctory *Proceedings of the Zoological Society of London*. Now another opportunity beckoned: publication in the same society's more prestigious *Transactions*. With this in mind, Owen began arranging for drawings to be made, and in one of several personal disagreements that would distinguish the history of this bone, he may have done so without Rule's permission. There was also a lack of unanimity in zoological circles regarding Owen's conclusions. The Publication Committee of the Zoological Society noted that because the bone was not a fossil it may have therefore come from an existing bird, and it seemed unlikely that a 'heavier and more sluggish species' than the ostrich could have gone undetected by the naturalists now combing New Zealand. In a note headed ominously 'The Bone', Owen recorded the misgivings of his senior colleagues, in particular those of Nicholas Aylward Vigors, Fellow of the Royal Society and founding member of the Zoological Society. His concern was that such a large bird as proposed by Owen had only the 'Island [*sic*] of New Zealand' for its habitat. By comparison, other known large birds of the scale of the ostrich, rhea, emu and cassowary had the extensive landmasses of Africa, South America, Australia and New Guinea respectively to range over. Owen had a reservation of his own regarding the alleged habitat of such a 'huge terrestrial bird'. There was no precedent for such a monster, for the largest New Zealand land bird then known to England was the barely pheasant-sized *Apteryx* (later known as the kiwi), and even its existence had not yet been confirmed.[20]

In another surviving scribbled note Owen provided a different perspective on the bone, now hinting that Rule's vital part in the story would be short-lived. He was reduced to 'the individual who brought the bone', and apparently when he 'penetrated the sanction of the head of natural history' – George Gray – at the British Museum, he was told they 'did not deal in old bones'. Fortunately, for both Owen and science at large, the 'Bringer of the Bone' was redirected to the Royal College of Surgeons.[21]

It was a calculated risk, suggesting the existence of such a huge creature on the basis of a single small fragment. But it was a reflection of both Owen's rapidly rising scientific reputation and his deductive powers that the Publication Committee of the Zoological Society agreed to publish his next paper on the subject in their *Transactions*, but with the proviso – at Owen's urging – that full responsibility for it rested with him.[22] This paper was probably available in September 1840, and was later bound into the next volume of the *Transactions* which appeared in 1842.[23]

Owen entitled his second publication on the bone, 'Notice of a Fragment of the Femur of a Gigantic Bird of New Zealand'. He had no further evidence to go on, but affirmed his original conclusion and gave a lengthy description of the process that led to it. He recorded that the bone had been brought to him for examination by Mr Rule, and repeated the previous references to the 'Eagle' and 'Movie'. But on this occasion he suggested that '[t]he first and most obvious idea of the nature of the bone would probably be that it belonged to the human species', but its size – nearly double the circumference of the femur of an ordinary-sized man – had ruled this out. Following up the possibility that it had belonged to a domestic animal as introduced into New Zealand by settlers, for 'food or draught', Owen explained how the femur differed from those of a whole procession of likely suspects, beginning with the ox. He also worked through the camel, llama, kangaroo and dog, admitting the 'improbability of their ever having found their way to the island [*sic*] of New Zealand'. In so doing he inadvertently displayed his unfamiliarity with both the geography of that country and the fact that its original Polynesian settlers had introduced a species of dog. Leaving no bone unturned, he also considered and discounted the grisly bear, lion, other large cats and the hog.

Although Owen felt that Rule's fragment suggested a bird with proportions similar to those of a dodo, he did not yet consign it to the same fate as that late bird from Mauritius. He cautioned that it would be premature to pronounce it extinct in view of the 'partially explored state' of New Zealand. He expressed his hope that this paper might accelerate the bird's discovery if it was still in existence, or at least stimulate the collection of the remaining parts of the skeleton if it was not. To this end Owen later claimed he had arranged for an extra one hundred copies of his latest paper to be printed for distribution throughout New Zealand.[24] In his hunt for 'confirmatory' evidence he was able to take advantage of an extensive number of contacts, for his circle of colleagues based around Lincoln's Inn Fields included lawyers and Oxbridge-educated clerics, several of whom were either associated with current plans to colonise New Zealand or about to make the trip themselves. If their luggage for the new colony did not include copies of his published papers, they are likely to have sailed with personal instructions from Owen to keep an eye out for anything that might be of benefit to science and, incidentally, his own illustrious career.

But the most significant advance in the second paper on Rule's bone was the inclusion of a plate of four full-size drawings by Bavarian-born artist George Scharf, who had arrived in London in 1816. These captured every crack, crevice and reticulate indentation of the humble fragment, with one interior view clearly showing the lamello-cellular structure or cancellous texture Rule claimed to have brought to the doubting Owen's attention.

While the learned societies of London awaited further evidence of this alleged ornithological oddity from the Antipodes, John Rule hoped he might find a buyer for the bone that had sparked the intrigue. Meanwhile, unbeknown to them all, there had been a rash of related developments back in New Zealand.

# 2

## THE FLAX FACTOR
### From Poverty Bay to Lincoln's Inn Fields

*The Bone they have a tradition that it belonged to a Bird of the Eagle kind, but which has become extinct–*

JOHN WILLIAMS HARRIS, 1837[1]

Dr John Rule's delivery of a bone fragment to the Hunterian Museum in London represented the final leg of a 12-thousand-mile journey that began over two years earlier. Sea travel was hazardous enough in the 1830s, so it would have been a cruel twist of fate if after weathering the lengthy voyage from New Zealand the bone was denied an audience with Richard Owen. That it got as far as England in the first instance was due to the activities of commercial traders based in the new penal colony of New South Wales. Their main interest was hardly bones – and certainly not those of extinct birds – but rather the economic possibilities of a plant with spiky leaves.

Lieutenant James Cook, RN, was the second European to rediscover New Zealand, 127 years after Dutchman Abel Tasman had sighted the country and mistakenly believed it might be connected to South America. On 7 October 1769 Cook promised a gallon of rum and naming rights to the first man on the

*Endeavour* to sight land, which he believed was near. The honour, if not all of the rum, went to twelve-year-old surgeon's boy Nicholas Young. The next day Cook set foot in New Zealand for the first time, on the site of present-day Gisborne, on the east coast of the North Island. But his first encounters with local Maori were marred by confrontations, and because he was also unable to obtain necessary provisions for his ship he decided to name the area Poverty Bay.

The *Endeavour*'s naturalists Joseph Banks and Daniel Solander went ashore at Poverty Bay on what would qualify as New Zealand's first botanical expedition, collecting many previously unknown specimens. By the time the ship left the country it had on board some 360 species of plants, including a relative of the European flax. Cook returned to New Zealand on two subsequent voyages, and on the second, on the *Resolution* in 1772, father and son German naturalists Johann and Georg Forster found several plants in flower, among them the local flax. They gave it the scientific name *Phormium tenax*, a reference to the traditional Maori use of its fibrous leaves for basket-making: *phormium* being Greek for 'basket', and *tenax* Latin for 'strong'. Although he recognised it might be difficult to transplant on account of its fine seeds, Johann Forster saw a great future for the local flax, suggesting that if established in Europe it might soon become 'one of the most usefull materials for manufactories in Linnen, Canvas, & rope'.[2] Also promising was the relative availability of this remarkable material, for when the *Adventurer*, accompanying the *Resolution*, called at Tolaga Bay, some 45 km north of Poverty Bay, astronomer William Bayley recorded the friendliness of the natives and their willingness to sell anything, fine flax included, except their greenstone adzes and ornaments.[3]

European interest in the South Pacific increased in the wake of Cook's voyages. It was suggested that a settlement there would be compensation for Britain's recent loss of its American colonies, as well as a handy solution to the problem of its overcrowded gaols. The British Government duly selected Botany Bay on the coast of New South Wales for their penal outpost, and the first convicts were dispatched in 1788. This distant location also had the advantage of its relative proximity to New Zealand, only six days' sail to the east, where large stocks of flax were now viewed as having great potential for the British Navy. A 10-inch circumference cable made of New Zealand flax was said to be as strong as one of nearly twice that dimension made of European hemp, while the southern plant also produced superior canvas and was a substitute for the finest silk. It was anticipated that this plentiful resource would be of great economic importance to the new settlement of Sydney, providing clothing for convicts and other settlers, as well as being a valuable export item.[4] There is a case for seeing proximity to New Zealand flax – and timber – as a key factor in the founding of European Australia,[5] and that new settlement now looked at the commercial possibilities of resources across the Tasman.

Flax from New Zealand soon gained the attention of enterprising traders in Sydney. One of the earliest of these was Simeon Lord, transported in 1791, allegedly for stealing 10d worth of cotton goods. When pardoned two years later he began a succession of businesses, before deciding to return to the commodities that had brought him to Australia in the first instance: textiles. He announced plans for a flax business in New Zealand, which would involve the procuring and preparing of raw material and the making of cordage and canvas at Port William, on the north-east coast of Stewart Island, the most southern of New Zealand's three main islands. He envisaged an expansive empire, largely dependent on the co-operation of local Maori, believing this contact would have the additional social benefit of putting these labourers on the road to 'progressive civilization', encouraging their learning of English as well as such trades as blacksmithing, shipbuilding and sawmilling.[6]

Another who entertained similar dreams was Robert Williams,

owner of a Sydney rope-works. In 1813 he crossed the Tasman on the *Perseverance*, and saw 'low land as far as the eye could discern' covered with flax. Believing the New Zealand plant could be processed at less cost than any other equivalent in the world, he sent samples to England for testing. And while the report from the British Navy's Chatham Rope Yard was positive, it cautioned that the specimens submitted were too small for a conclusive judgement.[7] Undeterred, Williams transplanted cuttings from flax roots, and found that leaves taken from the plant in the morning could be manufactured into cordage the same day. Like Lord, he saw a huge Maori labour pool just waiting to be put to work in a factory. With 'one month's instruction' he imagined they would be able to process 'immense quantities' of product. As a social spin-off, he believed this industry would soon improve his Maori workers' 'barbarous habits of life'.

Such was the faith in flax that in 1813 two 'machines' were brought over from Sydney for separating by boiling the fibrous parts of the flax leaves. But the technology proved ineffective, and was further hampered by the lack of suitable firewood. So much for the industrial revolution in the Antipodes, for it was found that Maori women, using the traditional manual method of scraping flax leaves with sharpened mussel shells, were far more efficient. As a result, most of the fibre taken to Sydney was hand-dressed in this manner, and it was not until the 1860s that a suitable machine – or 'stripper' – was devised to release the valuable white fibre from the leaf.[8]

It was not just the fibrous nature of *Phormium tenax* that lured Sydney businessmen to New Zealand. They also came to cut the country's massive stocks of timber, and to catch the whales and clobber the seals that frequented its coasts. Gangs of sealers were deposited in isolated corners of the country, and frequently for lengthier periods than they had anticipated, so were in a good position to observe some of the more unusual features of its flora and fauna. In 1822 the *Snapper* arrived back in Sydney, its cargo from New Zealand including flax and a large collection of bird skins, among them 'many new kinds'.[9] In addition to the usual commodities – including the now rapidly declining stocks of seals – other curiosities from New Zealand were now making their way to the outside world.

One of these was due to the intervention of Joseph Barrow Montefiore, who was born in London in 1803 and spent his early years with the family agricultural business in the West Indies. Back in London he hankered after the outdoor life, and when the British Government opened up the colony of New South Wales by granting tracts of land to men with capital, he applied. He arrived in New South Wales in 1829, taking up farming and breeding livestock, as well as trading as Montefiore Brothers, merchants, in Sydney's O'Connell Street. The following year he made a business trip to New Zealand, stopping at Kawhia on the West Coast of the North Island, where he appointed an agent to barter with the local Maori. But further south at Kapiti Island the hostility of the local Maori convinced Montefiore to abandon any further exploration of that country and he returned to Sydney. He maintained his New Zealand business interests, and in 1831 dispatched an agent to establish a flax-trading business on the opposite – eastern – side of the North Island. His man, John Williams Harris, was based at Poverty Bay, a 10-kilometre diameter semicircular bight whose southern extremity was marked by the white cliffs of the headland named Young Nicks Head in honour of the *Endeavour's* surgeon's boy.[10]

John Williams Harris was born in Cornwall in 1808, and served in the Royal Navy before travelling to Australia. He was sent over to Poverty Bay from Sydney with two assistants and sufficient goods to establish his first trading station. These included four bales of woollens, nine cases of muskets, eight cases of ironmongery, 32 casks of powder, and other quantities of hardware, oil, pipes and tobacco, rum and sundry items. Harris also made the first purchase of land in Poverty Bay, a property of a little over one acre, on a site now occupied by Gisborne, and acquired from Maori owners for 29 pounds of powder, one axe, 48 pipes and 6 pounds of tobacco. It was here Harris stored his goods, in a building described as being 'of the same unsubstantial character as those which the natives occupied'.[11] He later built a more imposing house and store, and thus stocked and sheltered he became the region's first permanent trader, and in the terminology of the day was described as a 'flax-factor'.[12]

In 1832 a flax establishment − presumably Harris's − was listed as having 'five or six white men resident', but the northern shore of Poverty Bay was sparsely populated at that time. There was a pa on Tuamotu Island at the north-eastern end of the bay, opposite Young Nicks Head, but it was now smaller than the one Cook had observed there in 1769. There was only one palisaded village in the vicinity of the site of present-day Gisborne, which in 1841 was reported as having a population of about 100. Further to the south-west were the remains of a large pa, which had been built as a refuge for local Maori in case of invasion by Waikato tribes. There were also various fortifications, including a pa which contained numerous whare (sleeping quarters) and other structures. For many years Harris was the region's most prominent Pakeha − non-Maori − citizen, and his occupation gave him much influence and standing among the local people. He had married Tukura, a Maori woman of considerable rank, and his business prospered, for by 1841 his property was occupied by a six-roomed cottage, a two-storey trading store and other wooden buildings.[13]

In 1835 Harris bought a farm, his land setting him back 250 pounds of powder, 10 pairs of trousers, 10 duck frocks, a d.b. (presumably double-barrel) gun, 76 yards of calico, 10 shirts and nine boxes of percussion caps.[14] In so doing he is said to have laid the foundations of agriculture in Poverty Bay, but this rather overlooks the pre-European history of the region. The agricultural expertise of East Coast Maori was observed by men on the

*Endeavour* in 1769, Banks noting well-tilled land with plantations laid out in regular lines, and patches fenced with reeds '[p]laced one by another so that scarce a mouse could creep through'.[15] In 1835, Joel Polack, another traveller and trader who was about to play a part in the story of New Zealand's mysterious bird, paid his first visit to the district and saw 'the most fertile land that may be imagined'. He also noted that much of it lay unused because Maori feared being caught out in their fields by their enemies.[16] At around this time, when the demand for dressed flax began to dwindle, Maori here turned to the large-scale cultivation of maize and potatoes, and the breeding of pigs, all commodities which would pass through the local trader's operation.

John Harris has also been described as both the 'founder of Poverty Bay'[17] and the 'father of shore-whaling' in that district. He established its first whaling station alongside his trading post in 1837 after a trip to Sydney. He took over bundles of bone and casks of oil retrieved from a whale cast ashore at Poverty Bay, along with the usual flax mats, maize, pork, pigs and ham, and in Sydney arranged for the shipment back across the Tasman of rum, gin, tobacco, tea, pork, beef, peas and empty casks.[18]

On his voyage to Sydney Harris took with him a small piece of bone, a fragment that had been recently discovered in a river bed and brought to his attention. In Sydney he decided to pass it and several other items on to his uncle, who happened to be Dr John Rule. The pair had arranged to meet, but because of other commitments and an earlier sailing of his ship back to Poverty Bay, Harris was unable to keep the appointment. In one sense this change of plans was fortunate, because Harris left his uncle a note, a copy of which is in the collection of The Natural History Museum, London, and provides evidence of the journey of the bone. Without this, even less would now be known about the origins of the specimen.

In his apologetic note, dated 28 February 1837, Harris provided brief descriptions of the five items he had left for his uncle:[19]

... first, the Stone Club – called by the natives of New Zealand Batoo Powato, from batoo to strike or kill, and powatu a stone. Carved box waka Wakero, from waka a canoe, and wakiro to carve. The two Wooden Clubs are called Maepe, I do not know from whence derived. The Bone they have a tradition that it belonged to a Bird of the Eagle kind, but which has become extinct – they call it 'A Movie'. They are found buried in the banks of Rivers.

This was the sum total of information transmitted by Harris to Rule, and three years later most of it would appear, almost word-perfect, in the opening paragraph of a paper published in the *Transactions of the Zoological Society of London* under the name of Richard Owen.[20]

Harris returned to Poverty Bay, but there is no record of any further contact with his uncle relating to the items he had left for him in Sydney. It had been hoped that much valuable information relating to the origins of European settlement in Poverty Bay – and, perhaps, background on the discovery of the bone fragment – would be found in a book compiled by Harris himself. But when this book was shown to a writer in 1926 – who published its main features in the *Gisborne Times* – there were disappointingly few references to events with which Harris had been associated. There was little mention of the establishment of the local whaling industry and the arrival of missionaries – significantly, Harris had been on good terms with the Rev. William Williams – and none concerning the bone he had taken to Sydney. It seems this was a later book, compiled by Harris's elder son, and the original was lost when the family home was destroyed by fire in 1868.[21]

By 1851 Harris was a settler of relatively substantial means, his property portfolio of five wooden buildings representing about one-quarter of those in the district. But around this time Tukura died, and Harris's business would soon suffer through competition. He remarried, later separating, and became increasingly depressed, committing suicide during a visit to Auckland in 1872.[22]

After the departure of his nephew, John Rule remained in Sydney until 1839 when personal circumstances dictated he return to England. In London he was determined to investigate the identity of the bone in his possession by comparing it with existing museum specimens. He was extremely thorough, reporting that his was 'larger in circumference than the largest in all the museums in London'. He gained access to both the 'public and private rooms of the British Museum' in Bloomsbury[23] and also inspected the collection at the Royal College of Surgeons, presumably not attracting the attention of the Hunterian Professor.[24] If the Zoological Society's collection was on his list, he would have had plenty to inspect there. Three years earlier, after having returned from his voyage on the *Beagle*, Charles Darwin had described the Society's museum as 'nearly full', and with upwards of a thousand unmounted specimens requiring attention.[25]

By comparing his bone with the femurs of certain large birds – and even with the thigh-bone of an ox – Rule found it was larger in circumference than all of them.[26] Convinced of its scientific significance, he also concluded that it might have some monetary value. With this in mind he approached the authorities at the British Museum, but found the trustees unimpressed by his offering. Fortunately, he encountered George Gray, who had been assistant in the zoology department since 1831, and would remain in charge of the Museum's bird collection until his death in 1872. In 1840 he would publish his *List of the Genera of Birds*, a book whose subsequent editions needed to expand rapidly to keep up with new discoveries by ornithologists. Gray must have been ever alert to the possibility of yet another species, but perhaps he did not regard Rule's as such. If Richard Owen's initial reaction was anything to go by,

Gray may have doubted that it was the bone of a bird, and thought it more likely to be related to the quadrupeds from South America that Darwin had recently deposited at the Royal College of Surgeons for identification. Gray therefore recommended that Rule take his bone back to Lincoln's Inn Fields and consult the authority on such matters, the redoubtable Richard Owen.

# 3

## TRADE AND EXCHANGE
### Petrifications and Ossified Parts

*I thus place myself in the arena of public opinion, regarding it as the
duty of every individual to add, to the best of his abilities, some
contribution towards the general treasury of knowledge; and, however
ill qualified for the task, he deserves well of his countrymen for the
intention . . .*

JOEL POLACK, 1838[1]

Like Joseph Montefiore, Joel Polack was born into a successful
Jewish family in London, in 1807. The son of artist Solomon
Polack, Joel was also a painter and exhibited – albeit one single
miniature – at the Royal Academy at the age of sixteen. But he
decided his prospects at the easel were not promising and settled
for a job at the War Office, which took him to South Africa and
Mauritius. After four years he left to travel to America, later
sailing to Sydney, Australia to join his brother Abraham in
business as a merchant and ships' chandler.[2] In 1831 he moved
to New Zealand, becoming its first Jewish settler, and worked as
a trader in the northern North Island. Before long he was based
in Kororareka, in the Bay of Islands, then the country's largest
European settlement and one with a reputation for lawlessness.

It would acquire the label 'Hell Hole of the Pacific', due largely to its numerous shorefront grog-shops which attracted deserted sailors and escaped convicts from across the Tasman.[3]

As a storekeeper Polack was one of the new 'merchant class' and he enjoyed good relations with the local Maori, a fact which didn't endear him to the settlement's less scrupulous operators. He also had some dealings with the local liquor business, claiming to have introduced 'New Zealand's first foreign manufacture' in 1835, by processing hops from Sydney in his own brewery. His stated aim was to stem the spread of deleterious spirits, which he claimed were consumed 'less probably from taste, than the want of invigorating substitute'. His 'preventive' venture proved successful, and after 'some little practice in quaffing' he noted that the local Maori were keen to exchange their baskets of potatoes and fish for 'pierian draughts of New Zealand beer.'[4]

Involvement in the Kororareka liquor trade was a certain recipe for trouble, and Polack managed to fall out with local innkeeper Benjamin Turner. Things came to a head in 1837 when the pair traded shots on Kororareka Beach, a confrontation claimed as New Zealand's first recorded duel. It was actually less of a formal affair than a running gun battle, in which Turner was wounded.[5] After this exchange, and perhaps because of it, Polack decided to lease his store and leave the country.

During the long voyage to England Polack found the time to reflect on certain of his recent exploits. In London these formed the basis of a book — his first — *New Zealand: Being a Narrative of Travels and Adventures During a Residence in that Country Between the Years 1831 and 1837*, published in 1838. In two volumes and approximately 175,000 words, the author aimed to 'excite attention' towards New Zealand by a statement of 'plain, unvarnished facts'. He claimed the 'strictest fidelity' but admitted he had been 'for many years sequestered from the society of literary men',[6] and now made up for his period of isolation with an extremely florid style of writing which has been described as having a 'tendency towards pedantry and prolixity'.[7] Among his many startling references to Maori – whom he compared to the ancient Britons 'in the days of Julius Caesar'[8] – was one to a chief of 'enormous muscular proportions' who could swallow the contents of a bucket full of 'cook's dripping and slush' and then ask for seconds.[9] If this seemed fanciful enough, Polack's book also had the distinction of containing the first published reference to the existence of a large flightless bird in New Zealand.

In 1835 Polack had chartered a small vessel to sail to Tolaga Bay on the East Coast of the North Island, enduring 'four successive gales from each quarter of the compass' on the way. Then, in retribution for the activities of previous Europeans to visit this part of the country. He suspected that both he and his ship were very nearly taken as a prize by the local natives.[10] It was during his stay in this district that Polack was shown 'petrifications' of the bones of large birds he believed to be 'wholly extinct', their tameness having made them easy prey for hungry natives.[11] He saw 'several large fossil ossifications', said to have been found at the base of the inland mountain of Hikurangi, and noted that local Maori enjoyed observing the European reaction to such things.[12] In language prescient of Richard Owen's first statements on John Rule's bone, Polack wrote: 'That a species of emu, or a bird of the genus Struthio, formerly existed in the [northern] island, I feel well assured . .'[13] Although confident these birds had long been 'extirpated', he had not given up hope, suggesting future ornithologists might be in for some pleasant surprises among the 'hidden mountain-gorges' and wilds of the South Island – which he curiously referred to as the 'Island of Victoria'.[14] Maori reports had convinced him that 'a species of struthio' was still to be found in its unexplored regions.[15] Moving into the twilight zone of ornithology, Polack recorded traditions 'current among the elder natives' of hair-covered and bird-like 'Atuas', (spirits or demons) which could overpower and devour unwary travellers in the bush.[16] Of the known and existing birds of New Zealand, Polack considered the most curious to be the 'kiwikiwi', or *Apteryx australis*, as included by John Gould in his book *The Birds of Australia, and the Adjacent Islands*. Polack repeated some of the author's other descriptions of this bird, but also included information he may have gleaned from personal experience – that the kiwi's flesh was 'worthless and tough'.[17] His list also included storm-birds and penguins, whose 'scaly' feathers suggested they were a link between birds and fish, or as he put it, the 'ornithological and piscivorous tribes'.[18]

As an author, Joel Polack hoped to contribute to 'the general treasury of knowledge' on New Zealand. That country was home to 'the most interesting race of uncivilised man', and one needing to be rescued from 'darkest barbarism and revolting superstition'.[19] This condition was reflected in the dramatic nature of the country itself, parts of which had been 'disembowelled by volcanic eruptions, and the coast cut into a thousand perforations by the lashing surges of the Pacific Ocean'. Cascades and waterfalls dashed from towering heights in this land 'rife with sublimity and grandeur'.[20]

Polack learned the language and established a personal rapport with Maori people, and whilst resident at Kororareka had first-hand experience of the undesirable effects of European influence on them. He praised the relentless and 'laudable' efforts of representatives of the Church Missionary Society, noting early difficulties caused by their determination not to supply Maori with European ammunition. But while Polack's own contact with missionaries had been marked by kindness and hospitality, certain unnamed brethren had been responsible for 'reprehensible conduct'. He identified six worthy individuals, one of whom was the Rev. William Williams, who would shortly play a major role in the discovery of the extinct bird whose remains had been seen by Polack. But his list of missionaries of character did not include William Colenso, another whose wide-ranging interests would extend to that same bird.[21]

This oversight may have been entirely innocuous, but Colenso later proved a difficult individual, earning the 'undying enmity' of his colleague William Williams.[22] Colenso took his ornithology seriously, and would not accept that Polack had seen those fossil bones as claimed. In fact, in what may have been an admission of his own likely response, Colenso suggested that if Polack had really seen such bones he would have 'grabbed' them. But the fractious missionary went much further – and lower – attempting to discredit the other by describing him as 'a Jew of the lowest grade'. He claimed he knew Polack well, even admitting he had often been in his 'rum store' on the Kororareka beach. Not stopping at anti-Semitic slurs, Colenso also stated that Polack had never even been to the East Coast or Poverty Bay and, further, that he could not – and indeed did not – write his books. He merely supplied 'rough materials' which the booksellers of London 'licked' into shape.[23]

Polack's comments on the local missionaries included a reference to one unmentioned individual who had conducted himself in so 'defective a manner' that he brought disgrace not only to his cause but to humanity itself. He was no doubt fulminating against the indiscretions of the Rev. William Yate, which led to his dismissal from the Church Missionary Society in England.[24] But Yate was hardly alone in straying from the path of righteousness, for even the Rev. Henry Williams, first on Polack's list of worthies, had a dispute with Bishop Selwyn in the mid 1840s over land claims by missionaries, and was dismissed – only to be later reinstated. Then there was William Colenso himself, who was sacked from the Church Missionary Society in 1852 for his sexual liaison with a Maori woman. It is just as well Polack was not able to allude to this incident in his book, for there is no telling how Colenso might have responded then.

During his time on the East Coast in 1835 Polack met a pair of hospitable European flax traders, who swapped several large hogs for tobacco, tomahawks, hoes and other 'trifles' he carried in a carpet bag.[25] There can be little doubt that Polack would have been aware of the wide-ranging John Harris, who had been in that district since 1831. Had they met they could have discussed the remains of extinct birds, but there was a point of geography that might have been of common interest as well. Polack recorded variations on Te Ika-A-Maui, the original Maori name for the North Island, among them 'E'Ainomawe' and 'Ai-no-maei' – which he translated as 'the begotten of Mawe'.[26] For amusement he also included a version used by French Captain Duperry which sounded suspiciously like Harris's mysterious 'Movie' bird: 'E'ka na mauvi'. Another curious French connection came in the form of Charles Philip Hippolytus de Thierry, better known as Baron de Thierry. As self-styled 'chief of New Zealand', he had arrived in 1837 with a bogus claim to the rights of sovereignty in the country. This prompted concerned British settlers to petition King William IV, and Polack included a full list of signatories in his book. He may have derived some satisfaction from the fact that among these settlers, church missionaries and others was his old adversary from Kororareka, who was not as handy with the pen as he was with the pistol, for he was listed as 'Benjamin Turner, his + mark'.[27]

Whilst in London Polack also became an active advocate for the organised settlement of New Zealand. A private company had been formed in 1838 to promote emigration to that country, and the British Government now decided to look into the matter. A Select Committee was convened by the House of Lords, and among the 18 witnesses who gave evidence were Joel Polack and Joseph Montefiore, both of whom were conveniently in England at the time.[28] Individuals giving 'evidence' did so with widely varying degrees of experience of New Zealand. Captain Robert FitzRoy, for example, had visited the Bay of Islands for ten days in 1835 during the course of his command of the *Beagle* and his five-year voyage in the company of naturalist Charles Darwin, and would return to that country as Governor from 1843–45.

In contrast, the Hon. Francis Baring, Member of the House of Commons, had never been to the country, but spoke 'from information'. Similarly, Lord Petre had been to several meetings of the New Zealand Association and had that country in mind for some of his sons, commenting: 'I have two of them very anxious for this sort of thing.' But any individuals thinking of emigrating may have had second thoughts when they heard the testimony of John Flatt, a missionary. He related how he had been stripped by the natives of everything but his horse, and presumed that was only because they had never seen such an animal before.[29]

Joel Polack made his first appearance before the Select Committee on 6 April 1838, describing himself as a Londoner who had been in New Zealand 'about six consecutive years'. He explained that he had spent much time among the Maori population, who freely told him their thoughts 'without disguise'. In his opinion the state of morals among the European population was 'decidedly bad', and he also mentioned that the conduct of 'one or two' of the missionaries had brought dishonour upon their names. Polack estimated the size of the entire country as 'nearly 100,000 square miles', which, if a wild guess, was remarkably accurate, the actual figure being 104,454. He pointed out that while there were ministers of the Church of England in New Zealand, there were none representing the 'Jewish persuasion'. He was one of only four Jews in the country, the only other one he was acquainted with being a cousin of fellow evidence-giver Montefiore. The Committee quizzed Polack about his own (five) property purchases in New Zealand, to which he explained the complications of land transactions. It seems even he was confused, for he was unable to confirm whether the size of the first lot of land he'd bought was 70 or 270 acres and whether he'd paid £15 or £25 for it. On other matters Polack was more certain, advising that venereal disease had been brought to the country by shipping, and that the natives there were 'exceedingly given' to intoxicating liquors.[30]

Four days later Polack was called back for a few more questions, and asked if he had kept a 'grog shop' in New Zealand. He replied that he had 'decidedly not', but admitted that 'as every person in the Bay of Islands did' he sold 'ardent Spirits'. He also pointed out that he was a wholesaler rather than retailer, so he never sold spirits to be consumed on his premises. Such comments may have raised the Committee's eyebrows, for it shortly asked Mr John Tawell, a surgeon who had recently been in New Zealand, if he had known the previously examined 'Mr. P -----'. Confirming that the latter's initials were J. S. P., Tawell admitted to having known him, and incidentally his brother, in New South Wales, but not in New Zealand. But when asked whether he should be 'designated as a respectable Man', Tawell confessed he was 'in possession of One or Two Facts of my own Knowledge which would make me disbelieve him on his Oath under any Circumstances'.[31] Polack also came under scrutiny outside the House of Lords when the *Times* newspaper referred to him as a 'worthy and wandering offshoot of the seed of Abraham'. On this occasion he had the satisfaction of successfully suing for libel, and claims he was awarded £100 damages by 'an enlightened jury'.[32]

The slights Polack endured in London were less personal attacks than popular prejudices. Had he consulted *Kidd's London Directory and Amusement Guide* he would have been warned of all that city's undesirable types, among them pawnbrokers, bailiffs, quack-doctors, swindlers, duffers, various types of tradesmen – including 'tricks of tradesmen' and 'religious tradesmen' – beggars, dress ladies and kept mistresses. The list included Jews, those 'troublers of Israel', who came in for special attention, with offensive comments extending even to matters of personal hygiene.[33]

In spite of strong support for colonisation expressed by witnesses, the House of Lords Select Committee's final report was somewhat inconclusive. Even as they deliberated, the New Zealand Company was preparing to sail with its first consignment of settlers for that country. Whilst he was in London Polack also arranged an auction of small lots of his own land in the country to which he was so energetically encouraging emigration. In so doing he may have been the first to subdivide into what became New Zealand's standard quarter-acre sections.[34] And encouraged by the favourable reception of his first book, he also set about producing another, *Manners and Customs of the New Zealanders*, published in 1840 for the benefit of intending emigrants. This second book also gave Polack a timely opportunity to clarify some of the allegations recently made about him, and he would have refuted the 'malicious attacks' made by 'calumniators' if his publishers had not advised against it. Nevertheless, Polack did respond to the accusations of one 'mendicious pamphleteer' that he had obtained land cheaply from the Maori, claiming he paid an 'equitable, fair and just price' for his five waterfront properties.[35] Despite his recent brush with the press, Polack preserved some degree of dignity by describing himself on the title page of his new book as 'J. S. Polack, Esq. Member of the Colonial Society of London'.

Unfortunately for ornithology, Polack's second book made no further reference to any 'osseous petrifications' or their like. He did however mention that many 'tribes' of New Zealand birds were already extinct, as victims of earlier human inhabitants and their own slothfulness.[36] In connection with the habitations of the gods, Polack referred to the mountain Hikurangi, near Waiapu on the East Cape, which natives never ascended alone, for it was the domain of the 'kiwikiwi' and other birds, as well as lizards. The most notable reference to birds in Polack's second book was in connection with their 'dawn chorus', in which his florid style took full flight: 'The feathered tribes . . . generally commence their concert before the dawn of day, which is swelled in sound by fresh choristers as they shake off the repose of night, continuing the chant until sunrise, when they separate, each to pursue his winged labours for the day, in providing sustenance for themselves and their young.'[37]

It was a feature of Polack's second book that, as if goaded by the *Times'* reference to Abraham, he missed few opportunities to embroider his already colourful descriptions of Maori culture with references to characters from the Old Testament. Abraham came up on the subject of polygamy, and to illustrate servitude in New Zealand Polack offered the example of Joseph, who had gone from being a slave to second only to the King of Egypt. This second book was also an opportunity to record how well the first had been received in England, for Polack claimed it was 'immediately noticed by nearly forty reviewers'. And because he knew none of these commentators personally he considered their criticisms 'unbiased'.[38]

In view of the growing awareness of New Zealand in England in 1838, we may wonder what effect Polack's first book had, and in particular whether its reference to an emu-like bird or fossil bones reached the attention of Richard Owen. There is no record, for example, of the library of the Zoological Society of London having had a copy of this book. But adventures in New Zealand were just the sort of thing carried by magazines of the day, and one source of 19th-century publications[39] lists articles on the *Apteryx* ('Bird or Reptile?') and other New Zealand birds, and three by J. Foster on Polack on New Zealand in 1831–37. These were published in the *Eclectic Review* and *Monthly Review*[40] and there is, of course, only a slight chance that any review of Polack's book might mention his 'petrifications' or 'fossil ossifications'. The search for any such references is made more difficult by the fact that the British Library's copies of the *Monthly Review* for the period 1831–44 were destroyed during the Second World War. While another source of 19th-century literature contains no specific references to Polack,[41] it does include at least one that would have been of interest to him: 'How to avoid fighting a duel', published in *Blackwood's Edinburgh Magazine* in 1840.

Joel Polack's first book did at least come to the attention of the New Zealand Company in London. In 1839 Company Secretary

John Ward published the first edition of *Information Relative to New-Zealand*, for those contemplating a new life in the spot 'most favoured by nature in the southern hemisphere'.[42] It listed previous publications on the subject of New Zealand, among them Polack's *Travels in New Zealand* [sic]. A copy of Ward's book in the Kinder Library at St John's Theological College, Auckland, was dated September 1839 by its first owner, so we know that Polack's book was circulating and being promoted at least a month prior to Richard Owen's encounter with John Rule's fragment of bone. And at least one writer presumes that Owen must have known of Polack's observations near Hikurangi at the time of the Zoological Society meeting in November 1839.[43] Meanwhile, Ward's book enjoyed a 'rapid' sale, necessitating a second edition in 1840, which continued to list Polack's book as essential reading for anyone considering life in New Zealand.

Both of Polack's books included engravings based on his own drawings, but unfortunately none of the large bird whose remains he saw. The closest he got to this was a view of a fortified Maori village 'near Poverty Bay',[44] a scene of much activity, with Maori chiefs urging on paddlers in a decorated war canoe with twin sails. In the background other vessels go about more mundane business, while a dozen or so thatched huts on shore suggest a settlement of a reasonable size. In his second book the author included a view of European sailing ships and a Maori canoe in front of a settlement, identified as 'Parramatta, Kororarika [sic] Bay, the residence and property of Mr Polack, Bay of Islands', which may have conveniently been the real estate the owner had put up for auction whilst in London.

In 1841 Polack returned to New Zealand, with at least one old score to settle. In a second duel with Turner, Polack took a bullet in the elbow while the other copped one in the cheek.[45] But there was a more serious confrontation brewing, for with the establishment of British government in New Zealand in 1840, many Maori in the northern North Island became concerned for their loss of power and authority. Ngapuhi chief Hone Heke became leader of the disaffected faction, and his first challenge to British authority was to cut down the flagpole above Kororareka, which he did on no less than four occasions. British troops were stationed in a local blockhouse, and in March 1845 Heke and his Ngapuhi ally Kawiti attacked. Residents were evacuated to Auckland, and the town was ransacked. By now Joel Polack had shifted his business operations to Auckland, but his own house back in Kororareka was destroyed – hardly surprising given that it had been used for the storage of British ammunition. By the end of the 1840s people were returning to the northern town, now renamed Russell in honour of Lord John Russell, Secretary of State for the Colonies. Now on the map of New Zealand, Lord John later became Prime Minister of Britain and also an elected trustee of the Hunterian Collection at the Royal College of Surgeons.

In both his books Polack suggested that the gum of the stately kauri tree might have commercial possibilities.[46] It certainly did, for the resin became a significant New Zealand export item until the development of synthetics began to replace its use in the manufacture of paints, polishes and linoleums in the early 1900s. In the late 1840s Polack exported a cargo of the gum to California, and in 1850 he also left New Zealand for new opportunities in the United States. He never returned, dying in San Francisco in 1882 at the age of 75.[47]

In 1840 Joel Polack described himself as 'an artist in Europe, a ship-chandler in Australia, a servant to the British Government in Africa' and 'a traveller for personal gratification in America'.[48] He has also been termed a 'cosmopolitan',[49] to which we might add adventurer, trader, duelist, writer and pioneer brewer. But perhaps most significantly, he was the first to publish a reference to the large flightless bird of New Zealand. Such a diversity of talents was not unusual at that time, and certainly handy for survival in the colonies. In the opinion of some – William

Colenso, in particular – Polack the writer was 'not always reliable'.[50] But for those who doubt his claim as a painter of miniature portraits who was also hung in the Royal Academy, the proof exists. According to the archives at the Royal Academy, he exhibited a single miniature painting, 'Portrait of a Gentleman', there in 1823, while his father Solomon showed no less than 57 paintings between 1790 and 1835. More tangible evidence of both Polacks' achievements can be found at the Victoria and Albert Museum in London, in a shallow glazed cabinet, fiche 22. Measuring 67 mm x 55 mm is an 1830 portrait entitled 'Unknown man'. This highly detailed portrait on ivory of a pale-complexioned young man gazing directly at the viewer is by Joel Samuel Polack, whose working period as an artist is recorded as 1823–30.[51] We might be tempted to suggest this is a self-portrait, but no known likenesses of Polack have survived for comparison. All we have is a pair of ink and wash drawings, dated 1845 or 1846, the one by Major Cyprian Bridge, and the other a copy by John Williams. Both offer only an outline of the slightly built trader Polack haggling with a trio of Maori over the price of a pig and some flax.[52]

We may never know for certain whether Joel Polack's references to 'osseous petrifications' and 'struthio' birds came to the attention of Richard Owen, or in any way influenced the statement he made at the Zoological Society of London meeting of 12 November 1839. Some might suggest it would have been in character for Owen to conveniently overlook any such inside information when he made his bold pronouncement. But we can be sure that Polack would have been interested in another find, some 30 years after his death. Around 1913 a local settler found the skeletons of five large birds in a cave in extremely rocky and bush-covered country near Mount Hikurangi. The birds were mostly complete and well-preserved, and all lying with their heads together at the end of the cave, suggesting that they may have been fleeing from an enemy or perhaps a fire.[53]

# 4

## WITH GOD ON THEIR SIDE
### Missionaries on the Path of the Moa

*. . . all agreed that it was a bird called a Moa; – that in general appearance it somewhat resembled an immense domestic cock, with the difference, however, of it having a 'face like a man' . . .*

WILLIAM COLENSO[1]

In June 1834 the Reverend William Yate left New Zealand after six and a half years of missionary work among the Maori people, and on the voyage back to England began writing an account of the work of his employer, the Church Missionary Society.[2] When published in London the following year, this unofficial history was greeted with disapproval by Yate's fellows, as would certain aspects of his personal conduct not long after.

Yate's book included a detailed description of New Zealand's 'most remarkable and curious bird', the kiwi. It certainly was curious, combining parts from several other animals. In size it compared to a three-month-old turkey, it had feathers like those of the emu and a beak like a curlew's. Its eyes were 'always blinking', and from its nostrils sprouted black hairs like the

49

whiskers of a cat. It had neither wings nor tail, but an acute sense of smell and short strong legs which provided for a fair turn of speed. According to Yate, kiwi were highly prized by the Maori, who used their skins for making into prestige garments, but he had only seen one such item during his time in New Zealand. And while there were few of the birds in the northern part of the North Island, there were said to be plenty near Hikurangi on the East Cape. And in noting also that the 'kiwi' there were larger than elsewhere, Yate may have been referring to another as yet unknown bird.

But if, like several of his fellow missionaries, Yate had plans to investigate this bird further, he would be disappointed. Back in Sydney in 1836 he was confronted by scandal, accused of having practised oral sex and mutual masturbation with young Maori men in the Bay of Islands. He returned to England in disgrace, where he was dismissed by the Church Missionary Society. He was appointed chaplain to a mission for seamen at Dover, and died there in 1877.[3]

With a few notable exceptions, missionaries toiled to bring enlightenment and salvation to the uncivilised souls of New Zealand, and several also played a leading role in the development of science in that country. Their primary mission was all part of a divine plan, set in motion when the first fleet was about to sail from England to New South Wales in 1787. At the last minute an appeal resulted in the appointment of a chaplain to accompany the convicts on their journey to banishment in the Antipodes. Samuel Marsden, later New Zealand's first missionary, interpreted this decision as evidence of an all-wise and merciful God who enabled 'men from the dregs of society' – indeed, the very 'scrapings of jails, hulks and prisons' – to provide employment for His servants.[4] Britain's social outcasts now had purpose to their lives, if only to provide work for clerical brethren.

Samuel Marsden was born in Yorkshire in 1765, and after studying at Magdalene College, Cambridge, was appointed assistant chaplain to the colony of New South Wales, reaching Sydney Cove in 1794. When he met Maori visiting from New Zealand he became concerned for their spiritual welfare, seeing their minds as 'a rich soil that had never been cultivated, and only wanted the proper means of improvement to render them fit to rank with civilized nations'. Believing the race was under the influence of the Prince of Darkness and in urgent need of the Gospel, Marsden went to London in 1807 to explain the situation to the Secretary of the Church Missionary Society. It was decided that three missionaries would be sent out with Marsden on his return, but no clergymen rose to this vocational challenge, presumably deterred by reports of the wild and lawless nature of life in those southern parts.

Despite initial setbacks, Marsden pressed on with his plans and in November 1814 made the first of seven trips across the Tasman to that country in need.[5] He established a mission station at Rangihoua, in the Bay of Islands, and others followed, all controlled from his base at Parramatta, near Sydney. These were now staffed by qualified missionaries, who travelled widely to tend their far-flung flocks. When Marsden visited Waimate and Omapere in the far north of New Zealand in 1815, his was the first significant journey of inland exploration by a European in the country, and set a pattern for the missionary explorers who followed.[6]

Christianity was not all Marsden had hoped to introduce to New Zealand. Dedicated also to agricultural improvement, he took over a range of livestock with him. And if he was not so interested in the existing fauna of the country, he did introduce one who was. When Marsden's brig *Active* left Port Jackson for New Zealand in 1814, it carried animals, three missionaries and their families, as well as John Liddiard Nicholas, who had recently arrived in Australia from England and befriended Marsden. While other passengers pursued more spiritual aims, Nicholas observed New Zealand's natural history, publishing his account on his return to England in 1817. He recorded that the feathers used in garments worn by Maori chiefs were similar to those of the emu of 'New Holland' (Australia) and therefore presumed the existence of a species of cassowary in New Zealand. However, the bird eluded him, not least because it was in fact the kiwi. Nevertheless, the mistaken Nicholas may have been the first to suggest New Zealand was the home of large flightless birds.[7]

In 1823 the Church Missionary Society in England appointed retired Royal Navy midshipman Henry Williams to take over the running of its three stations now operating in New Zealand's Bay of Islands. Shortly thereafter, he was joined by his younger brother William, who had been born in Nottingham in 1800. After completing an apprenticeship to a surgeon, William Williams entered Magdalen Hall (later Hertford House), Oxford, as a Church Missionary Society trainee. While he was there he encountered the colourful and influential Reverend William Buckland, the first reader in the new science of geology, a discipline – or rather a personal passion – he referred to as 'undergroundology'.[8] Buckland was a colourful character with a high profile, and, according to one source, Williams attended his lectures.[9] He gained a BA in classics, and arrived in New Zealand in 1825 to take charge of the English boys school at Paihia. In addition to the mission station, this was also the site of the first church to be erected in New Zealand, a structure constructed of raupo – a New Zealand variety of bulrush – in 1823.

To better spread the word in New Zealand, the Church Missionary Society in London now sought the services of a suitable printer. Cornishman William Colenso was appointed, and he arrived in the Bay of Islands on 30 December 1834. His first challenge was to get his press and printing equipment ashore at Paihia, which lacked a suitable jetty. Colenso overcame the problem by means of a platform set on two canoes lashed together, a makeshift vessel reminiscent of those used by early Polynesian voyagers. An empty classroom would be his printery, but after unpacking his crates Colenso discovered his employers back in London had overlooked several printing essentials, including paper. He was forced to improvise, using the services of a local joiner to make vital wooden components, and a stonemason to dress a pair of basalt boulders for his imposing stones. But getting the press rolling was not Colenso's only task, for he also acted as local doctor, provided classes for adult Maori, attended daily prayers in the chapel and took several church services on Sundays, as well as dealing with much of his own housekeeping and the other demands of an isolated community during a time of tribal conflicts. In addition, he somehow managed to find time to study the Maori language.[10]

With the press finally ready to roll, Colenso could begin his first job. Naturally, it would be from the Scriptures: *The Epistles to the Ephesians and the Philippians*, which William Williams had just finished translating into Maori. On 17 February 1835, in front of a crowd of curious onlookers, Colenso pulled proofs of the first book printed in New Zealand, using oddments of paper collected from around the community. He then turned to other Scriptures in Maori, also translated mostly by Williams, and by the end of 1837 had finished printing the New Testament. The missionary committee was so impressed with the results that they granted the printer and his editor leave of absence from mission duties, whereupon the pair undertook a voyage to the East Cape.[11]

If William Williams had already encountered one of science's more colourful characters in the form of William Buckland, both he and William Colenso also met one who would become its most notable. On 21 December 1835 the *Beagle* nosed into the Bay of Islands, carrying Charles Darwin. The young naturalist's first impressions of New Zealand were less than favourable, and hardly improved when he stepped ashore. He found much of the landscape impassable, 'uninhabited useless country', and the native people more savage than civilised. At Waimate, home of William Williams and other missionaries, Darwin saw a small patch of transplanted England: fine crops of wheat and barley, fields of potato and clover, large gardens of every fruit and vegetable imaginable, a happy mixture of pigs and poultry, a blacksmith's forge, and a mill attended by a Maori who was dramatically whitened by a dusting of flour. Such sights reassured Darwin, for he could see beyond 'what Englishmen could effect' to the future prospects for the country. While it was a land of 'cannibalism ... and all atrocious crimes', he never saw a merrier group than at William Williams' house one evening. The Reverend also took Darwin to a nearby forest of kauri trees, where he viewed one specimen measuring 31 feet in circumference. But this was the only local giant Darwin saw, and it was not until after his return to England and introduction to Richard Owen that he learned of the 'several species of that gigantic genus of birds, the Deinornis [sic]' that had once lived in that country.[12]

On 30 December 1835 the *Beagle* set sail from the Bay of Islands for Port Jackson, and Darwin believed he was not alone in being glad to leave New Zealand. It was, he wrote, 'not a pleasant place', but he reserved his harshest comments for its English residents, describing the greater part of them as 'the very refuse of society'. In a quarter of a century he would become the most famous scientist of his time, but in the meantime he put his faith in religion, for as he sailed away he reflected on New Zealand's one bright spot: 'Waimate, with its Christian inhabitants.'[13]

In early 1838 William Colenso hung up his printer's apron and put aside his composing stick. He and William Williams sailed for the East Coast, following a route already familiar to John Harris, Joel Polack and other traders, crossing the Bay of Plenty and landing at Cape Runaway. From there they walked overland to Hicks Bay and on to Waiapu, a 'thickly inhabited' locality some 30 km south-west from the East Cape. It was here that Colenso first heard about a singularly large and unusual creature, a monstrous animal which resembled an immense domestic cock except it had a 'face like a man'. Local Maori informed him it inhabited a cavern on the precipitous side of a mountain, where it reportedly lived on air and was guarded by a pair of huge tuatara (lizards) while it slept. Further, anyone who dared to approach the haunt of this remarkable creature would face certain death. Colenso also learned the name of the mountain – Whakapunake, a 975-metre limestone scarp that lay some 130 km to the south – which was said to be the home of this creature, the sole survivor of its race. But while there was general acceptance of its existence, Colenso could find no one who could claim to have actually seen it – or, as he quaintly put it, 'had ocular demonstration of it'. The closest the locals had dared venture to it was their occasional discovery of very large bones, which they had used in the manufacture of fishing lures. Colenso saw no such bones at this stage, and neither could he persuade anyone to take him to the creature's lair. The two missionaries returned to the Bay of Islands without any tangible evidence of this beast, but Colenso had at least taken note of its name, and was the first European to record its identity as the 'moa'.[14]

Nearly a year later, in January 1839, William Williams returned to the East Coast, now accompanied by the Reverend Richard Taylor. He was a Yorkshireman who had graduated from Queen's College, Cambridge, in 1828 and taken up an appointment as a missionary in New Zealand in 1839. The Reverends Williams and Taylor sailed to Hicks Bay and from there travelled overland to Waiapu. While he was there Taylor noticed a large bone fragment in the thatched ceiling of a Maori chief's hut, and traded a quantity of tobacco for it.[15] When he returned to the Bay of Islands he showed it to Colenso, who described it as '(what appeared to be) a part of a fossil toe (or rather claw) of some gigantic bird of former days'.[16] But years later, Colenso was much less complimentary about Taylor's find, now referring to it as a 'so-called "toe"' resembling a 'bit of water-worn and rolled Obsidian', or dark volcanic glass.[17]

There are discrepancies between Taylor's later recollections and his journal of the day, but the latter records that whilst he and Williams were in Poverty Bay in 1839 they dined at the house of trader John Harris. Harris now gave Taylor a bone – or bones – of a bird which he described as 'larger even than an ostrich'.[18] But if he told the missionaries about the bone he had passed on to his uncle John Rule in Sydney just over two years earlier, he was probably unaware that the soon-to-be famous specimen would shortly be – if not already – on its way to Richard Owen in London.[19]

Taylor learned from local Maori of a valley near Tokomaru, some 37 km north of Tolaga Bay, where 'the great bird moa was said to exist'. They also told him that they dreaded to hear its call, regarding it as 'a certain portend [sic] of death'.[20] As far as the missionaries were concerned, it was more an omen of controversy, reflecting the growing prestige and competitiveness now associated with the discovery of that extinct bird. While Taylor later made an extraordinary claim in that regard, he had no argument with the name, 'Moa'. It was not until 1843 when he moved from the Bay of Islands to Wanganui and travelled around the South Taranaki coastline – an area also then known as Waimate – that he first heard the word. Here he discovered local Maori 'totally ignorant' of the name given to the bird on the opposite side of the North Island – 'Tarepo'.[21]

During the summer of 1841–42, William Colenso made a second visit to the East Coast. On this occasion he was told by some local Maori – perhaps now fortified by Christian training – that they had visited Whakapunake, the alleged mountain home of the moa. Unfortunately their expedition was fruitless, for it provided no sign of the creature, or even its cavern or lizard-guards. But, at last, Colenso was able to get his hands on some bones – seven in all: five femora (plural of femur), one tibia (shinbone) and one unidentified – which Maori had found in the bed of the Waiapu River and declared to be from the moa.[22] Colenso then left Waiapu and travelled south and met up with

William Williams, now stationed at the Turanga mission station on the site where Gisborne now stands. Since their last meeting Williams had also obtained a moa bone, his being a nearly whole tibia some 18 inches long.[23] Williams now needed an informed opinion on the subject, and knew just who to consult: Buckland, at his old university. Colenso left a pair of his femora to accompany the planned shipment, and continued southwards in search of the moa. He hoped to find bones at Whakapunake, but was again disappointed. If it was any consolation, close examination convinced him that no such giant animal could live on this mountain anyway, it being 'huge, table-topped, and lofty' and covered with 'primeval forests of gloomy pines'.[24] At Te Reinga, a village at the foot of the mountain, locals confirmed that the moa lived there but – the usual story – no one had actually seen it. Colenso also learned about moa living on a mountain further inland, but when he arrived there fifteen days later a close inspection with the aid of his pocket telescope also revealed nothing.[25] Hearing that bones were commonly found exposed in river banks after heavy rains, Colenso now offered 'large rewards' for any that might now be delivered to William Williams at Poverty Bay, and then returned to the Bay of Islands.

Colenso's trip back to his home in the north was the first European crossing of the interior of the North Island, a vast zigzag undertaking which took him on 'byways never before trodden by the pakeha'.[26] He did it on foot and by canoe, assisted by Maori guides who occasionally did a lot more than just showing the way and carrying his supplies. On one occasion a porter slipped down a steep slope and collided with a tree, and, although unhurt, the box of books he was carrying was smashed to pieces.[27] When it came to fording a stream on the lip of a 20-foot cataract, the nervous missionary was carried across the slippery rocks on the back of an obliging and sure-footed guide.[28] The journey traversed grassy plains and primeval bush, tangled bracken and fern-covered ridges, mountain ranges and gloomy river gorges.

At times there was: 'Fern, fern, nothing but dry dusty fern all around.'[29] On the way Colenso was constantly collecting botanical specimens, but did not forget his main responsibilities. In fact, he was forced to question the basis of his own religion when his path coincided with that of a Roman Catholic priest, Father Baty. The two met, and on Christmas Eve, 1841 there ensued 'violent controversy between the two representatives of the gospel of peace', much to the entertainment and confusion of their Maori audiences. Shortly afterwards Colenso felt duty-bound to visit another village to reconvert souls who had recently received the annoying Father Baty.[30]

But there was some gratifying evidence of progress. At Whakawhitira, a Maori village above the Waiapu River, Colenso reported many changes since his visit three years earlier in the company of William Williams. Then, the inhabitants were 'living in the grossest hour of heathenism'; none could read or knew 'anything of an hereafter'. Now, nearly 700 assembled for service in a chapel they had built of the bark of the totara tree, measuring some 80 feet by 40. Colenso also witnessed the improving hand of God in the insects he gathered. He noted that in many 'their colour was assimilated to that of the plant on which they lived', natural adaptation which he interpreted as 'a beautiful display of the Divine Wisdom.'[31]

Colenso's route took him via Lake Waikaremoana, Ruatahuna and Te Whaiti, over the Kaingaroa Plains and the Whirinaki and Rangitaiki Rivers, to Rerewhakaaitu, Rotomahana, Lake Okareka – which he crossed by canoe – and the Te Ngae Anglican mission station near the eastern shoreline of Lake Rotorua. From there he went north to Tauranga, and down the 'sullen' Waipa River to Ngaruawahia, where he took advantage of the '[p]owerful rolling waters of the Waikato'. Downstream, about a mile from the sea, Colenso called at the mission station of the Rev. Robert Maunsell. The latter had just completed compiling his *Maori Grammar*, first published in 1842, so the two men spent a day discussing the original language of New Zealand.[32] Colenso then walked north along the west coast to the Manukau Harbour, camping for the night among the sandhills. From there he canoed across to Otahuhu, improvising sails from blankets to take advantage of a breeze, and on to the west side of the Waitemata Harbour. The long road north went via the Kaipara Harbour, and the travellers were forced to cross the Waipu River estuary by placing their clothes on a log and pushing it before them as they swam.[33] While Colenso was being carried across the Horahora River his porter was bitten by a small fish, whereupon the inquisitive missionary immediately jumped into the water to secure this specimen for science.[34] Finally, on 22 February 1842, seven weeks after beginning the homeward leg of his journey, Colenso arrived back in Paihia.

Meanwhile, back in Poverty Bay, Colenso's request for specimens was having the desired effect. The promise of payment had produced something of a 'bone rush,'[35] with some bones of an enormous size and in a good state of preservation being brought in by bounty-hunters. Williams now had the bones of nearly 30 birds, including a tibia measuring 2 feet 10 inches in length, and two others at 2 feet 6 inches. He and Colenso were now both busy in the service of the moa, in different parts of the North Island and some 500 km apart. They had the disadvantage of pursuing their inquiries in an isolated country, and without colleagues experienced in such matters. Scientific convention demanded that their specimens be submitted to experts and their findings published, and to do so the missionary scientists needed to look further afield. For his part, Williams bundled up bones for the attention of William Buckland. Mindful of the hazards of sea travel, he divided his valuable cargo into two crates which were carried separately by boat from Turanga to Port Nicholson (Wellington), and then shipped via Sydney to London.

Up north at Paihia, Colenso was working on his contribution to the world of knowledge. This was a scientific paper, and there was a likely publication fairly close at hand, thanks to Arctic explorer and Battle of Trafalgar veteran, Sir John Franklin. He was appointed Lieutenant-Governor of Van Diemen's Land (later Tasmania) in 1837, and upon his arrival in Hobart had quickly set about promoting cultural activities in what had been a penal colony. He founded the Tasmanian Natural History Society, which encouraged citizens to study aspects of their 'adopted country'. In 1845 Sir John's own commitment to scientific inquiry would take him on another, and on this occasion ill-fated, expedition to the Arctic, in search of the North-West Passage. While scientific exploration was very much a male domain, Sir John's wife Lady Jane Franklin also promoted the pursuit of knowledge. Her support extended to sending a botanical microscope to William Colenso to assist his studies.[36]

Members of the Tasmanian Natural History Society were busy men, their everyday duties allowing little time for scientific pursuits. Although further disadvantaged by the distance from Europe, their remote corner of the world had much in the way of unknown plants and animals to arouse the curiosity of their peers back Home. Their organisation – which became the first scientific Royal Society to be established outside Britain – maintained two classes of members: resident and corresponding. The former, as published in the first issue of the Society's *Journal*, included several from New Zealand, among them William Colenso and Richard Taylor. From further afield came the Rev. Professor Buckland (Oxford), John Gould, F. L. S. (London), Captain Stokes, R. N., of HMS *Beagle*, Joseph Hooker of HMS *Erebus*, along with others from that ship and its companion vessel on its Antarctic expedition, HMS *Terror*. The list also included another individual who was already playing an important part in providing specimens for British science, and would soon be in a position to deal with the moa, His Excellency Captain Grey, of Adelaide, South Australia.[37]

From New Zealand, Richard Taylor and William Colenso took advantage of the *Tasmanian Journal of Natural Science* to publish the results of their own scientific investigations. The first of these to appear was 'The Bulrush Caterpillar', under the name of the Rev. R. Taylor of Waimati [*sic*]. His subject — now known as the vegetable caterpillar — was a plant found at the root of rata trees which took possession of the body of an insect. Taylor observed that this curious plant — of which he had examined at least a hundred — was used by the Maori both as food and a colouring medium, when burnt, for tattooing. He also regarded it as a case of a retrograde step in nature, for instead of 'rising to the higher order of the butterfly, and soaring to the skies', this caterpillar became part of a plant and remained rooted to the soil.[38]

In Paihia in early 1842, William Colenso was preparing to submit two papers of his own to the *Tasmanian Journal of Natural Science*. One of these was entitled 'Description of some new Ferns lately discovered in New Zealand', and referred to botanical specimens collected in '[h]umid places in the dense forest' on the east coast of the North Island. This was where he was also on the lookout for material for his other paper, which recalled the events of the summer of 1838, and his hunt for a 'certain monstrous animal'. In his second sentence Colenso would divulge its name, for 'all agreed that it was called a *Moa*'.[39] In describing the tuatara which guarded the beast's haunt Colenso illustrated the difficulties of the New Zealand scientist, for he had no reference books to assist his identification of this reptile. He had earlier kept one alive for nearly three months, but was unable to induce it to feed and it was now reduced to a spirit specimen.[40]

With the delivery of a large number of specimens, William Williams was able to attempt a rough assembly of the leg and thigh bones of a moa, and estimate that the complete bird had a height of some 14 to 16 feet. On this basis Colenso described it as an 'enormous feathered monster', worthy of being classed with two other recently discovered giants: William Buckland's Megalosaurus and French naturalist Georges Cuvier's Mastodon.[41]

Turning to the problem of where the moa stood in the natural order of things, Colenso again admitted the difficulty of making deductions without 'any known osteologic specimens for comparison, or any scientific books for reference'.[42] The latter were, of course, the very things Richard Owen enjoyed in abundance at the Hunterian Museum, and the reason new and problematic specimens were entrusted to him for identification. Even so, Colenso placed the moa in the Struthio family, just as Joel Polack had done some five years earlier. Colenso had had 'several' kiwi in his possession on previous occasions and now wished he had one at hand to check his suspicion that it was related to the moa. Looking further afield, he also suggested the moa might be the 'connecting link' between the *Rhea* of the Straits of Magellan, the *Dromiceius* of New Holland, the *Casuarius* of the Indian Archipelago and the *Apteryx* of New Zealand.[43]

For his paper on the moa, Colenso produced drawings of a selection of bones. He regretted that these did not include 'the largest and most perfect' specimens, because they had been collected and sent to England by William Williams after Colenso had left Poverty Bay for Paihia. Instead, he drew a 'nearly perfect tibia' which was only four inches shorter, along with a 13-inch femur, a 10-inch tarsus, and fragments of the pelvis and dorsal vertebrae.[44]

On 1 May 1842 Colenso finished his paper on the moa and submitted it, along with his other on ferns, to the Tasmanian Society. They were 'laid before' the meeting of 3 August and Colenso was advised that the fern paper would be included in the next (fifth) issue of the *Journal*, which was already under way, while the other (moa) paper would be reserved for the succeeding number.[45] The editor also referred to the challenge of printing illustrations 'properly', and hoped that Colenso would be happy with the results.[46] It seems that technical difficulties associated with reproducing drawings influenced the Society's decision to publish Colenso's fern paper first. This was most unfortunate from the moa's point of view because other factors would now delay the appearance of that paper until 1846.[47] A rival scientific body supported by the new Governor, Sir John Eardley-Wilmot, had serious ramifications for both the membership and the publications programme of the original Tasmanian Society.

Such were the frustrations of the antipodean naturalist. After several extremely fruitful months of collecting and writing, Colenso was suddenly thwarted. The only chance of publishing locally was now denied him. But like William Williams, Colenso had a good contact in England, and so sent on a copy of his now delayed paper. That well-connected contact knew exactly what to do, and passed 'An Account of Some Enormous Fossil Bones of an Unknown Species of the Class Aves, lately discovered in New Zealand' on to the obvious authority – Richard Owen.

# 5

## DISSENSION IN THE RANKS
### Conflicting Claims among the Missionaries

*. . . but I think I may justly claim to have been the first discoverer of the Moa.*

RICHARD TAYLOR[1]

A s John Rule and his bone were making their way towards England, another expedition that would have implications for the moa was preparing to travel in the opposite direction. Promoted by the British Association for the Advancement of Science, and under the command of Captain James Clark Ross, the *Erebus* and *Terror* set sail from Chatham in late September 1839 to explore the seas around Antarctica. The two ships spent three winters in the southern ocean, and when conditions made Antarctic exploration impossible they sheltered at Tasmania, the Falkland Islands and New Zealand. From August to November 1841 they were based at the Bay of Islands, where the assistant surgeon and botanist on board the *Terror*, 22-year-old Joseph Dalton Hooker, made the acquaintance of William Colenso. The pair went on botanical collecting expeditions, striking up a lifelong friendship and corresponding for the next half century.

Joseph Hooker returned to Britain in late 1843 and began work on publications of his new specimens. When the first volume of his *Flora Novae-Zelandiae* appeared a decade later, William Colenso, 'the foremost New Zealand botanical explorer', was prominent in the dedication. In Hooker's opinion he was one of five individuals who stood 'pre-eminent as indefatigable collectors and explorers'.[2] But when it came to botany in the colonies, Hooker believed much more public education was needed, and recommended that labelled collections of dried plants be made available for study in schools and libraries. As a juror at the Great Exhibition of 1851 he had seen large botanical collections which were 'all but valueless' because of their exhibitors' lack of knowledge, and this was particularly true of those sent from New Zealand.[3]

Prominent among those doing their bit for colonial science was William Colenso, who had also received some systematic training and essential equipment from Allan Cunningham, the New South Wales government botanist who visited the Bay of Islands in 1838. Cunningham may have been less impressed by Colenso's unsystematic habit of stuffing specimens down his shirt front whilst out collecting however, a technique we can only assume did not extend to the several species of native stinging nettles.[4]

During his stay in New Zealand Joseph Hooker regularly sent specimens back to the Royal Society, the Admiralty and to his own father, Sir William Hooker, who had recently been appointed the first director of the Royal Botanic Gardens at Kew. The elder Hooker now also received a consignment from his son's new friend, William Colenso, consisting of a scientific paper on several New Zealand ferns, along with an 'admirable memoir on the fossil bones of a bird allied to the ostrich' and specimens of its bones.[5] He passed the moa material on to his colleague, Richard Owen, who arranged for its publication.

Colenso's paper on the moa appeared in the *Annals of Natural History* in August 1844,[6] to be followed over a year later by its much delayed appearance in the *Tasmanian Journal of Natural Science*.

Nearly half a century later Colenso wrote that he was unaware of any differences between the two versions,[7] but there were in fact several. Most strikingly, the *Annals of Natural History* version explained in detail how the monstrous moa of Whakapunake dispatched its victims, using a technique similar to that by which criminals were 'summarily punished in the dominions of the native Indian princes, by the trampling of an elephant'. Moreover, this was a feat at which 'the celebrated Moa was quite expert'.[8] Perhaps this gem was added to Colenso's text in England to complement the information on the two 'immense tuataras' which guarded the sleeping moa. Colenso identified them as belonging to the class *Reptilia* and order *Sauria*, while Maori spoke of two other species, one with a forked tail and the other, a swamp-dweller, capable of growing to six feet and as thick as a man's thigh.[9] As for smaller saurians, Colenso claimed to have at least six distinct species in his collection, and mentioned that they were 'objects of superstitious dread' to the Maori. The tuatara was said to be eaten by one or two tribes, who as a result were often 'spoken contemptuously of by their countrymen'.[10]

Turning to the age of the moa, Colenso was surprised to find it had been largely forgotten in the otherwise extensive oral tradition of the Maori, and he was therefore convinced the bird had become extinct either prior to or soon after the arrival of the Maori in New Zealand.[11] As for the name, he noted that in several groups of islands in the Pacific, the domestic cock was known as the 'Moa'. It may have also reflected an ancient memory of the cassowary of the Malay Archipelago, which was perhaps located on the migratory route of the ancestors of the Maori.[12] Small pieces of moa bone were used by the Maori for fishing lures, and, speaking as a missionary, Colenso could report that their hooks were often made from human bone 'previous to the introduction of the Gospel among the New Zealanders'.[13]

The most significant difference between the two versions of Colenso's paper was the matter of nomenclature and the moa's place in the animal kingdom. He identified the recognised members of the Struthionidae family, which then consisted of six genera: *Struthio* (ostrich of Africa), *Casuarius* (cassowary of the East Indies, New Guinea and Australia), *Dromiceius* (emu of Australia), *Rhea* (rhea – then known as nandu – of South America), *Didus* (dodo of Mauritius) and *Apteryx* (kiwi of New Zealand). In his original text, as printed in the *Tasmanian Journal of Natural Science*, Colenso had put forward a seventh genus: the 'Moa of New Zealand'. But if he imagined that 'Moa' might form the basis of the generic name for this newly discovered bird, he would be disappointed. By the time Colenso's paper had appeared in the *Annals of Natural History* there had been significant developments in London, and an asterisk now alerted the reader to a footnote. This advised that Professor Owen had recently assigned the generic name *Dinornis* to this 'monster bird', and that 'no less than five species [had been] distinguished'.[14] And when his paper finally appeared in the Tasmanian publication, Colenso would notice it was immediately followed by an item based on a letter from Professor Buckland to Sir John Franklin. This gave some background to the above development, and as a result the great bird of New Zealand will stand as representative of: – *Deinornis Novae Zelandiae*, Owen [sic]'.[15] The Hunterian Professor had therefore put his authoritative stamp on the moa from New Zealand.

William Colenso and William Williams had now furnished their scientific masters in England with both specimens and a scientific paper, and their contributions to moa research were largely complete for the time being. Williams remained based at the Turanga mission station in Poverty Bay from 1840 until 1865, and continued his overland journeys, but was now more interested in the training of a Maori pastorate than collecting further moa bones. Colenso continued his itinerant ways, leaving mission work to enter local politics, but in 1879 he was eventually drawn back to the subject of the large extinct bird.

Meanwhile, after visiting the East Coast with William

Williams in 1839, the Rev. Richard Taylor was put in charge of the Church Missionary Society's school at Waimate (now Waimate North) in the Bay of Islands. He remained there until 1843, when he was appointed to a mission station at Wanganui, on the West Coast of the North Island. From this base he made regular visits to all the villages in the district. On one such journey around the coast, at the mouth of the Waingongoro Stream seven kilometres to the west of present-day Hawera, he found a fragment of bone which reminded him of the one he had picked up at Waiapu. Further investigation revealed a sandy plain 'covered with a number of little mounds, entirely composed of Moa bones'. It was, as he put it, a 'regular necropolis of the race'.[16] Changing his priorities, Taylor emptied a box of his supplies and filled it with moa bones, much to the amazement of his Maori entourage. He intended to return to this rich source of specimens, but by the time he did – over three years later – it had been well picked over by another enthusiastic collector named Walter Mantell. He was another who sent regular shipments of bones back to England, which invariably found their way to Richard Owen.

Although deprived of the best specimens at Waingongoro, Taylor's inspections of the coastal sandhills rarely failed to produce remains of the moa. On one occasion he found what usually eluded the moa hunter, a complete skull, partly embedded in clay soil. It was 'in a most friable state', and after careful excavation he placed it in the crown of his hat for safety. Unfortunately, his horse reared and sent him sprawling, shattering his rare treasure. Nevertheless, he managed to gather up the fragments and send them on to Professor Owen. In 1866 Taylor made his final visit to the Waingongoro site, in the company of Sir George Grey – now serving his second term as Governor of New Zealand – and found the hillocks were again thickly covered with bones, the result of wind erosion. A series of old Maori ovens provided dramatic proof that took the moa out of the mythological past, for these remains of meals past included relics of the bird alongside stone flakes and polished adzes. The men busied themselves gathering specimens, and Taylor was amused by the sight of His Excellency, 'grubbing up the old ashes, and carefully selecting what he thought worth carrying away'.[17]

At Waingongoro, Taylor was told by local Maori that the moa was as large as a horse, which he took as 'sure proof' that none of them had ever seen the bird.[18] Interestingly, this comparison of the equine species with flightless birds was also noted by Charles Darwin. At Maldonaldo in Uruguay in 1832, during the *Beagle* voyage, the naturalist observed a Tierra del Fuegian boy's first encounter with an ostrich, which he described as 'bird all same horse!'[19]

In 1855 Richard Taylor published *Te Ika a Maui, or New Zealand and its Inhabitants*, a detailed account of the country's Maori life, geology and natural history. Unfortunately, in the light of later developments, he did not include his own alleged contributions to moa discovery, restricting his comments to the demise of this 'gigantic ostrich', a victim of climate changes rather than human intervention. With its loss the Maori became 'greatly pinched for food', but Taylor saw the arrival of Europeans as a tribute to God's own providence, for they brought 'fresh means of sustenance'.[20]

Then, in 1872, as if driven by a zeal to convert, Taylor revealed his version of the 'first discovery' of the moa. This bold and belated claim related to the trip he and William Williams made to Waiapu in 1839. He now insisted that when retrieving the fragment of bone from the ceiling of a Maori chief's hut, he had immediately noticed its 'cancellated structure' – terminology suspiciously similar to that used by Richard Owen in his first paper on the moa. Taylor then claimed that Williams dismissed the suggestion that this might be the bone of a bird – a surprising reaction in view of the fact that this large creature had been the subject of much investigation and discussion when Williams and

Colenso had visited the East Coast the previous year. After parting with a 'little tobacco', Taylor acquired the bone and 'in the beginning of 1839' sent it on to Richard Owen. He now claimed that some months later another such bone was obtained by a 'sailor' in the same district, and it was this one that John Rule had taken to England. Although it had reached Owen 'some time' before his own specimen, Taylor felt he was 'the first discoverer of the Moa'. His claim, of course, had no basis; firstly, because Rule's bone was already in Sydney by February 1837 — when John Harris wrote the accompanying note — and secondly, the individual who took Taylor's bone to England could not have done so until 1846.[21]

Richard Taylor claimed his bone was delivered to Richard Owen by one Everard Home, which not only destroys his 'argument' but provides another curious moa connection. Home's father, also Everard was born in 1756, and at the age of 16 became a pupil of surgeon John Hunter, living at his house and personal museum in Leicester Square. Hunter's sprawling collection now included material brought back by James Cook and Joseph Banks from the *Endeavour* voyage, and was being sorted and catalogued by another young pupil, Edward Jenner, who later achieved fame for discovering an inoculation against smallpox. Upon Hunter's death in 1793, one of Everard Home's duties as executor was to convince the Government to buy the late surgeon's collection. That achieved, Home's career went into the ascendant, especially after he had performed a successful operation on George III's fifth son, the Duke of Cumberland. By the time of the opening of the Hunterian Museum in 1813 Home was Master of the Royal College of Surgeons, and would shortly be knighted. But ten years later he divulged a dreadful secret, that it was he who had destroyed John Hunter's original manuscripts — in all likelihood because he wanted to take credit for the other's work in the advancement of his own career. The Hunterian Conservator, William Clift, was left to sort out this mess, a challenge made easier with the appointment of an assistant – and future son-in-law – Richard Owen, in 1827.

Meanwhile, Sir Everard Home's son – the younger Everard – had opted for a naval career. He commanded the frigate *North Star* during the so-called Opium War with China, and from there sailed to Sydney in 1843. This was a time of mounting tension in New Zealand, and Home transported soldiers across the Tasman to support European settlers against the threat of Ngati Toa chief Te Rauparaha in the central part of the country. Shortly after, there was trouble up north when Ngapuhi chief Hone Heke began threatening British authority and later attacked the town. In early April 1845 Home arrived at the Bay of Islands in the *North Star* to evacuate to Auckland a group of residents assembled by missionary Henry Williams, and it was presumably then – or on the ship's return to the Bay of Islands – that Home met Richard Taylor and received the bone the latter intended for the attention of Richard Owen. Home then took the *North Star* to Sydney, and from there sailed for Portsmouth, arriving in August 1846. If the bone was indeed on board, it would have been delivered to the Hunterian Museum some seven years later than Taylor claimed.[22]

In 1844, while Kororareka was under threat, William Colenso was ordained deacon and offered a post in a new mission station in Hawke's Bay. He took responsibility for a huge parish, and the need to travel provided opportunities for him to escape from an unhappy marriage. In 1852 a sexual liaison with a Maori woman led to his expulsion from the church, and he then entered local politics in Hawke's Bay. He also became a prolific writer of scientific papers, and returned to the moa when he presented a lengthy paper on the subject to the Hawke's Bay Philosophical Institute in 1878. He believed the bird worthy of being 'classed among the animal wonders of the world', but it was hardly known in Hawke's Bay. He tabled some bones – noting that very few had been found during his 35 years in the district – along with photographs of mounted specimens, suspecting that many in his audience had never seen or heard of such things. And while he had published – twice – on the subject, he feared he had the only copy of his paper in the Colony; what's more, he had been unable to procure one 'at any price' in London. Colenso then read the text of his original paper – now 36 years old – and followed it with an update entitled: 'What I have gleaned since'. Among other things, recent research had convinced him of the antiquity of the moa, the bird having died out 'in the time of Noah'.[23]

Colenso's paper was an insight into the working methods of the European moa hunter. From 1838 until 1842 he felt that no man could have done more than he did in the quest for the bird. In doing so he admitted that few men had better opportunities, for Church Missionary Society work required him to travel widely in isolated areas of the country. He was convinced that it was by his exhaustive efforts that the Maori population got to know of the moa as having been 'a *real* (or common) bird'. This roving missionary showed them illustrations of the known struthious birds from a copy of *Rees' Cyclopaedia*, and it was testament to his dedication that this weighty volume was carried on his excursions.[24] Among its 250 pages of engravings – 'by the most distinguished artists'[25] – were two of struthious birds, although none from New Zealand. He therefore would have had to explain the object of his inquiry by showing these images of the rhea, cassowary, ostrich and dodo, the last-mentioned being useful for a lesson on extinction. Colenso felt that certain information on the moa collected from the Maori by other individuals – including even Governor George Grey – may even have originated from his own encyclopaedia. Thanks to his investigations he believed the moa had become a regular subject of conversation among the Maori, 'who then had little of a novel nature to talk over'.[26]

As for Richard Taylor's claim to have been first to discover moa remains, Colenso wondered why this had not been made public earlier than 1872. He pointed out that both men were early elected members of the Tasmanian Natural History Society, and Taylor was 'well-known not to be backward in writing of everything'. Reflecting on his own extensive Missionary Society activities during that period – which included the printing of over 74,100 books between 1835 and 1840[27] – Colenso retorted that 'while at the North [Taylor] had plenty of time to call his own'. But he resisted dealing with his colleague's outrageous claims, putting them down to memory failure and inattention to detail,[28] in sharp contrast to the attitude he took some 15 years later when he vehemently dismissed Joel Polack's writings.[29] Colenso concluded his extensive 1879 paper by announcing that he had 'finished [his] work' on the moa and – somewhat prematurely – would 'not again write on this topic'.[30]

But twelve years after he had expounded 'fully and exhaustively' on the moa and supposed he had 'quite done with it', Colenso was at it again.[31] In 1891 he was moved to respond to suggestions that claims made on his behalf by English geologist Gideon Mantell, the father of Walter, were a slight on Richard Owen. Three years after his first pronouncement on the moa, the Hunterian Professor had received the confirmation he was after (the subject of the next chapter). This latest squabble involving Colenso further highlighted the isolation of the antipodean scientist, for the article in question had appeared in an English journal a full forty years earlier,[32] and he had never seen it. Better late than never, he now sourced a copy from the General Assembly Library in Wellington. It was fulsome in its praise of Owen, noting that once 'all criticisms and misgivings as to the original audacious induction from the fragment of the supposed marrow-bone being thus quashed, there remained only attempts at detraction from the merit of discovery'. This was a reference to Mantell's claim for Colenso's priority in the discovery of the struthious character of the moa. While the article did not aim to detract from Colenso's observations, it certainly cast doubt on Mantell's judgement. Colenso took up the cause, feeling the other had been hit 'very hard' and unjustifiably so, for he had hardly ignored or even slighted Owen's achievements.[33] Indeed, one of Mantell's own papers on fossil bird remains was unstinting in its praise of Owen's 1839 identification – now confirmed – calling it: 'One of the most interesting palaeontological discoveries of our times.'

In fact, in his 1848 paper,[34] Gideon Mantell could have hardly been more complimentary about Richard Owen's achievement: 'And if I were required to select from the numerous and important deductions of palaeontology the one which of all others presents the most striking and triumphant instance of the sagacious application of the principles of the correlation of organic structure enunciated by the illustrious Cuvier – the one that may be regarded as the *experimentum crucis* of the Cuverian philosophy – I would unhesitatingly adduce the interpretation of this fragment of bone. I know not among all the marvels which palaeontology has revealed to us a more brilliant example of successful philosophical induction – the felicitous prediction of genius enlightened by profound scientific knowledge.'

It was surely clear enough: Gideon Mantell, now a vice president of the Geological Society, had given Owen the highest possible scientific praise for being 'the first to announce to the world at Home his great discovery'. Equally, Mantell was justified in his carefully and lengthily worded claim that Colenso was 'the first observer that investigated the nature of the fossil remains with due care and the requisite scientific knowledge (having determined the struthious affinities of the birds to which the bones belonged, and pointed out their remarkable characters, ere any intelligence could have reached him of the results of Professor Owen's examination of the specimens transmitted to this country)'.[35] In other words, all Mantell was saying was that Colenso was the first to do so 'out here in the Antipodes'.[36] It was Colenso's view that Mantell had no wish to upset Owen, claiming that the 'utmost kindness, disinterestedness, and liberality' had long existed between them.[37] This was yet another indication of New Zealand's isolation, for far from being a pair of cordial colleagues, a bitter professional feud existed between Owen and Mantell, with the Professor frequently getting the better of his rival.

At the heart of this criticism of Mantell was whether Colenso had access to any of Owen's papers on the moa prior to publishing his own. Although Owen claimed that one hundred extra copies of his first *Transactions* memoir, 'Notice of a Fragment of the Femur of a Gigantic Bird of New Zealand', were 'distributed to every quarter of the Islands of New Zealand where attention to such evidences was likely to be attracted',[38] there is confusion as to exactly which paper was sent, and when. The issue is whether it reached New Zealand before 1 May 1842, the date Colenso completed his own paper on the moa which he submitted to the Tasmanian Natural History Society. For his part, Colenso could 'positively affirm' that not only did he never see Owen's first memoir, he never once heard of it, and neither did he know of any other 'resident in New Zealand' who had seen it. It was hardly surprising if Colenso was unaware of such a publication, for – as he pointed out – he was living in a part of the country where 'no moa-remains had ever been found' and where even 'the name was unknown'. Also, because the Bay of Islands was distant from the main centres, with only irregular communication by 'small coasting-vessels', he was not in the habit of receiving the sort of material Owen claimed to have circulated.[39]

On the other hand, Colenso was not entirely cut off from the outside world. In an earlier publication he had listed some of the 'scientific gentlemen' he met whilst living near the anchorage at 'the Bay'. These included members of Captain James Clark Ross's Antarctic Expedition, the United States Exploring Expedition under Commander Wilkes, and several French ships of war and discovery. He 'saw and knew them all' and they 'zealously sought after any remains of the *Moa*', but without success. There were also many traders now digging for kauri gum who were likely to have found moa remains, 'but they, too, got none!'[40] And then there was Ernst Dieffenbach, the New Zealand Company naturalist based in this country during the period 1839–41. For a time he lived near Colenso in Paihia, and while the pair often discussed the moa and 'kindred matters', he too appeared not to have heard of Owen's first memoir. Similarly, Colenso had communicated with William Williams and stayed at his house in Poverty Bay, and was convinced that he also was unaware of the Professor's paper at that time.[41]

To further prove that he had been operating oblivious to developments in London, Colenso turned to the individuals whom Owen suggested had 'efficiently aided' the dispersal of his memoir – Colonel William Wakefield and J. R. Gowan of the New Zealand Company, Chief Justice Sir William Martin and Bishop Selwyn[42] – and proceeded to suggest why none could have done so. He quoted a letter from Wakefield to Gowen dated September 1843 which indicated that the former had only just heard of Owen's request for bones. Colenso also suggested that Martin had 'vastly too much of higher and more important public matters' than the moa to attend to, while Bishop Selwyn did not arrive in New Zealand until June 1842, the month after Colenso's paper was completed.[43]

When William Colenso finally had access to another of Owen's papers, his 1844 'Memoir on the Genus Dinornis' – which followed the arrival in Oxford of the bones sent by William Williams and confirmation of the existence of the moa – he would learn that the Professor gave Williams 'a just claim to share in the honour of the discovery of the *Dinornis*'. Owen did so because that 'zealous and successful Church Missionary' had been collecting and comparing the bird's osseous remains 'wholly unaware that its more immediate affinities had already been determined in England' – a reason that must have sounded very familiar to Colenso.[44] And if he was disappointed by suggestions that he had investigated the moa in full knowledge of Richard Owen's announcement of 1839, Colenso was also 'not a little surprised' to discover that William Williams had not even mentioned his contribution in the letter that accompanied the bones to Oxford. That consignment consisted of specimens that local Maori had delivered to Williams, as well as a pair of femora Colenso had carried from the East Cape, some sixty miles distant.[45]

In spite of any slights or oversights, William Colenso was the first dedicated moa investigator on the spot, and also the first to put his findings in writing, if not in print. Had his paper appeared in 1843 when it was received by the Tasmanian Society, he might also have been the first to publish the word 'moa', but that same year Ernst Dieffenbach included it in his *Travels in New Zealand*. Colenso has been termed the 'father' of scientific botanical study in this country.[46] He wrote more than one hundred published scientific papers, and in 1886 had the distinction of becoming the first New Zealander to be elected a Fellow of the Royal Society.[47] The indefatigable Colenso's interests were as wide-ranging as his cross-country journeys. At an 1894 meeting of the Hawke's Bay Philosophical Institute he was responsible for three of the five papers presented, on 'Animal Prodigies', greenstone and newly-discovered plants. At the close of that meeting a presentation was made to Colenso, the President describing him as a man of scientific attainments such as were recognised in Europe and America, but wondered whether he was as well known in the town where he had lived for half a century as 'the ordinary handicapper or jockey who rode in a race'. Making the obvious analogy, he commented that the 82-year-old Colenso was still treading the 'pathways of science with pleasure', and pursuing the 'hobby of his life'.[48]

If William John Colenso was a controversial character, it ran in the family. John William Colenso, a cousin three years his junior and also from Cornwall, was appointed first Bishop of Natal in 1853. Nine years later he published his doubts on the historical truth of Biblical events, convinced by his knowledge of geology that Noah's Flood could not have occurred as described in the Book of Genesis. Accused of heresy and rejected by the majority of his colleagues, he managed to continue as Bishop of Natal until his death in 1883. In a 'duo-biography' of the two Colensos, A. L. Rowse has noted their similarities. They were both frontiersmen ('the one working for the Zulus, the other for the Maoris') who were forthright, combative, utterly honest and altruistic, as well as being outspoken, tactless and completely irrepressible. Perhaps the main difference was that William – the Deacon – was 'even more dogmatic and forthright' than his cousin the Bishop.[49]

The New Zealand Colenso also had a penchant for the unusual. On one of his early journeys he discovered the mysterious bronze Tamil Bell, so-named because an embossed inscription suggests it originated from the Portuguese colony of Goa in India. According to information Colenso provided when he bequeathed the item to the Colonial Museum (now Te Papa Tongarewa Museum of New Zealand), the antique bell was found 'in the interior of the North Island' – probably Northland – in 1836.[50] Another of Colenso's sizeable contributions to local natural history and now secure in the Auckland Museum is a giant weta, its label noting that he 'captured' it in 1838. This gargantuan – whose species name *Deinacrida* has a common root with that of the moa – is the oldest New Zealand insect in the Museum's collection, and has the added distinction of having been displayed at the New Zealand Exhibition at Dunedin in 1865.[51]

From Poverty Bay William Williams had dispatched moa bones to the Rev. William Buckland in Oxford, while his colleague, Paihia-based William Colenso, sent both bones and a scientific paper to Sir William Hooker at the Royal Botanic Gardens at Kew. Upon receipt of their respective consignments, both gentlemen knew exactly what to do. When it came to the moa, all roads led to Richard Owen.

# 6

## GIVE THE MAN A BONE
### A Case of Severe Philosophical Induction

*As each bone of the feathered giant was taken out it was impossible to repress exclamations; but when the enormous tibia came within our grasp, it was flourished aloft with a shout of wonder and joy that made the Museum ring again.*[1]

Screened by foliage and lurking at a discreet distance, Richard Owen watched as with much circling and sniffing on the part of the male, the two giraffes began their preliminary formalities. He had undertaken to study how two of the largest attractions at the Zoological Society's Gardens at Regent's Park went about reproducing their species, and while it might seem unusual work for a comparative anatomist, there was none better qualified than Owen to do it. The animals he usually concerned himself with were dead – and in many cases had been so for several epochs – but these subjects were very much alive. He scribbled in his notebook and another scientific paper took shape, based on the 'circumstances of connexion' as observed on two occasions – on 18 March and 1 April 1838. Owen likened the 'mode and brief duration of the coitus' to that of the deer. It was mostly a silent operation, for 'only at the period of the rut' was any sound uttered, the male giraffe emitting a short guttural bleat like that of the stag. The observant naturalist also noted that after the second connection, the female showed 'no further disposition to receive the male'.[2]

These giraffes were among a succession of exotic animals that arrived in London in the early 19th century. The public turned out to marvel at such wonders of nature, but Richard Owen's interest was in more serious matters. On 10 April 1838 he explained to a meeting of the Zoological Society how that animal went about swallowing, using what he termed its 'organs of deglutition'. He had been puzzled by the giraffes' reluctance to produce vocal sounds – except, of course, during the time of coitus – and could now put this peculiarity down to the nature of the glottis, explaining with references to 'infundibular forms', 'valvular ridges' and 'mucous follicles'. Then, having dealt with the intimate workings of the tallest quadruped, he moved to his next topic, the anatomy of a small bird from New Zealand. Later, such lectures were the basis of scientific papers published in either the Society's *Proceedings* or *Transactions*. On this occasion there would be another positive outcome of Owen's field-work, when the Zoological Gardens were able to announce the birth of a baby giraffe.[3]

Born in 1804, Richard Owen was a Lancaster lad, the son of a successful West India merchant. His early schooldays were hardly promising, and his only interest was heraldry. We cannot be sure whether it was the attention to detail or the association with exotic beasts that fascinated him, but both would prove useful in later life. Owen contemplated joining the navy, but decided instead to go to Edinburgh University to study anatomy. He didn't complete his degree, and came to London with a piece of paper that proved far more useful – a letter of introduction to the influential surgeon John Abernethy. He appointed Owen prosector (dissector of cadavers) for his surgical lectures at St Bartholomew's Hospital, and when the prospect of permanent employment there did not eventuate, Owen reconsidered life at sea, as assistant surgeon on a ship. Abernethy intervened, and in the opinion of one commentator, 'the navy lost a good officer, and science gained one of her brightest ornaments'.[4] In 1826 Owen gained membership of the Royal College of Surgeons and set up in private practice in Lincoln's Inn Fields, but his interests were already moving towards comparative anatomy.[5]

Custody of John Hunter's personal collection of specimens had been assumed by the College, which was under increasing pressure to prepare and label these items for public exhibition. William Clift, who had been John Hunter's personal assistant, was appointed the first Conservator of the Hunterian Museum, which opened in 1813. It proved an instant attraction, receiving some extremely distinguished visitors – all of which heightened the need for a descriptive catalogue of its contents. The obvious person for the job was Sir Everard Home, Master of the College, who as Hunter's executor was in possession of the manuscripts that were the key to the collections. But time passed and nothing happened, and in 1823 Home made the astonishing admission that he had destroyed Hunter's manuscripts. An obvious motive was plagiarism, for Home had lifted large sections from Hunter's original work for inclusion in his own publications. In his defence, Home claimed – unconvincingly – that Hunter had wished his papers to be destroyed upon his death because they were unsuitable for public viewing. The zealous executor had disposed of nearly 90 per cent of the original manuscripts, and a visit to his house even revealed one – *Treatise on the Venereal Disease* – ignominiously lodged in the lavatory.[6] Other valuable medical manuscripts had no doubt already been flushed to a soggy fate in the Thames. The Trustees of the College faced a dilemma, and William Clift got on with cataloguing as best he could. He obviously needed assistance, and John Abernethy, now President of the Royal College, suggested Richard Owen for the job.

Contrary to Home's claims, Hunter did not of course intend

his manuscripts to be destroyed, for he considered them 'an integral and most important part of his collection'. Remarkably, while Home was, according to an official history, 'guilty of what was without doubt a serious indiscretion', he remained a Trustee of the Hunterian Collection and attended regular meetings until his death in 1832. Records from that time suggest fellow Trustees held him in high regard, appearing to have passed over any of his 'delinquencies'.[7]

Richard Owen was appointed Assistant Conservator at the Hunterian Museum in 1827. He quickly acquired a great reverence for the work of John Hunter, applying himself to the challenge of identifying, cataloguing and labelling each of the 13,000 specimens. He also found time to develop an interest in William Clift's only daughter, although at first he was considered an unsuitable suitor. But before long Owen's career prospects began a steady improvement, much to the satisfaction of his future in-laws, and so after an eight-year engagement the patient couple were permitted to marry.

Within a few years Owen had completed a reorganisation of the Hunterian Museum, an achievement which brought him much recognition. He had also made an important contact in French naturalist Georges Cuvier, who visited the Royal College of Surgeons in 1830 and invited Owen to study with him in Paris the following year. Cuvier was the pioneer of a new science that sought to establish the laws that governed the anatomy of creatures, and was based at the Museum National d'Histoire Naturelle, a centre of anatomical study that then had no equivalent in England. Back at Lincoln's Inn Fields Owen soon became known as the 'British Cuvier', but he would not languish for long in the shadow of the Frenchman. His reputation was on the rise, benefiting from a regular supply of new and challenging specimens dispatched for his attention from the far corners of the British Empire.

Such an item was presented to the Museum in 1831 by

George Bennett, who studied medicine at the Hunterian School of Medicine in the 1820s, where he struck up a friendship with Richard Owen. In 1829 he sailed on the trading vessel *Sophia* to Sydney as surgeon in charge of 92 convicts. After off-loading its human cargo the ship continued to the New Hebrides (now Vanuatu), where Bennett collected a live specimen of the rare pearly nautilus, previously known only by its shell.[8] According to the sailors, it was found floating on the surface near the ship and resembled 'a dead tortoiseshell cat'. It was seized with a boat-hook and, not surprisingly, found to be slightly damaged when taken aboard. Bennett's attention was drawn to the fact that it was 'possessing an inhabitant', which he immediately detached from its fractured shell. He then sketched it and placed it in spirits for delivery to the Hunterian Museum. In Owen's sure hands the animal became the subject of his 1832 'Memoir on the Pearly Nautilus', an elegant 60-page publication, with eight plates drawn up by the author.[9] It attracted much attention, and in the words of Thomas Huxley, 'placed its author at a bound in the front rank of anatomical monographers'.[10] Back in Australia, George Bennett later became the first Curator of the Australian Museum and supplied Owen with further specimens, including the paradoxical platypus and kangaroo. These natural curiosities stimulated further papers, simultaneously advancing Owen's career and drawing his attention to the scientific potential of those southern lands.

Richard Owen enjoyed a steady rise through the ranks at the Hunterian Museum. In 1842 the Assistant Conservator was appointed Joint Conservator with William Clift, and, following the latter's retirement, in 1852 Owen graduated to Senior Conservator.[11] In addition, in 1836 he was appointed the first Hunterian Professor of Comparative Anatomy and Physiology at the Royal College of Surgeons, a post that required him to deliver 24 annual lectures illustrative of the collections. Now known as Professor Owen, and always to be seen in his voluminous

Hunterian robe, he joined a select group of scientific men, hob-nobbing with royalty and other notables including Charles Dickens and Lord Tennyson, and discussing museum matters with Sir Robert Peel.

Thanks to Owen's industry, the long-awaited and much thwarted documentation of John Hunter's collection was now making good progress, and the catalogue of the physiological specimens appeared in five volumes between 1833 and 1840. Because of the loss of Hunter's originals, some 4000 items had to be identified by recourse to alternative documents, or by comparison with fresh dissections. To assist this process Owen took advantage of the availability of exotic animals that had died in the gardens of the Zoological Society. These unfortunately frequent events were nevertheless a good source of raw material, and among others things provided Owen with a freshly deceased orang-utan, beaver, crocodile, armadillo, kangaroo, tapir, pelican, cheetah, wombat, giraffe, dugong and wart-hog.

Another unusual subject now becoming familiar to Richard Owen was the kiwi. To satisfy the new fashion for sealskin in the 1790s, gangs of men had been sent to exploit the large colonies of those marine mammals around the coasts of southern New Zealand. One of these sealers is presumed to have been the agent for a different sort of skin that found its way across to Port Jackson, Sydney in 1811. It was from a strange and unknown bird, wing-less and with a long beak. It came into the possession of a ship's captain who was en route to China to take on a cargo of tea for England. When he reached London he did as John Rule would do some 27 years later, and took his unusual skin to the British Museum. There it came to the attention of Assistant Keeper George Shaw, who may have interpreted his job title a little too literally because the specimen became his personal property. However, Shaw recognised it as belonging to the same group as the emu and other flightless birds, and arranged for it to be illustrated in the 1813 edition of his *The Naturalist's Miscellany*. In its first published outing, this bird was given a stiff penguin-like pose, a result of the artist's faulty extrapolation of the body shape from the skin. Shaw named the bird *Apteryx australis*, the 'wingless bird of the south',[12] and when he died shortly afterwards his unique specimen was purchased by natural history collector Lord Stanley, later 13th Earl of Derby. It joined a private accumulation that included a museum of 20,000 specimens of quadrupeds, birds, eggs, reptiles and fish, and a menagerie of 345 head of mammalia and 1272 head of birds – not counting poultry.[13]

With the increasing number of European visitors to New Zealand, this curious bird could not elude science for much longer. Feathers were sighted, and a missionary reported a flightless bird known to the Maori as 'kivikivi', suggesting that this and the *Apteryx australis* were one and the same. A decade later, the sole skin specimen in England was re-examined and the bird was given a much more acceptable profile. In 1834 the Rev. William Yate in the Bay of Islands advised that he had recently kept such a bird alive in his possession for nearly a fortnight. Soon, information and actual specimens were reaching England. The Earl of Derby acquired another bird, and in 1838 presented a body and viscera to the Zoological Society, which also received examples from George Bennett and a Dr Logan. Naturally, all these came to the attention of Richard Owen.[14]

Now with sufficient material at hand, Owen began preparing his first scientific paper on a New Zealand bird.[15] He was also now able to introduce recently discovered information on the kiwi into his Hunterian lectures. In one of these in May 1838 he described the feeding habits of the long-beaked *Apteryx*, as observed by Dr Logan: 'It was seen to invariably poise itself before making a strike. It never missed hitting the exact spot where the worm lay.'[16] And for his introductory lecture in 1839, delivered on Tuesday 30 April, Owen drew attention to a few of the 'most novel and most interesting objects which have lately been contributed to the naturalist'. Amongst these the *Apteryx australis* stood 'pre-eminent in every point of view'. Owen explained why it was not like other birds, being 'utterly devoid of external wings'. Further, its feet were not constructed for swimming like the penguin, nor — he claimed — were they strong like those of the ostrich and adapted for speed. Further, the New Zealand bird was nocturnal, seeking for its food by night and 'lying hid during the day in the burrows it forms from beneath the surface of the earth, where the process of incubation is formed'.[17]

One of the challenges of the early-19th-century naturalist was the preservation of specimens. At a May 1839 meeting of the Zoological Society a Mr G. Smith exhibited several birds which had been successfully preserved, 'all the parts entire', by the injection of an 'antiseptic fluid'. Members were then advised that Smith had taken out a patent on his fluid. Also discussed that evening was another new kiwi specimen, preserved in acetic acid and presented to the Society by Allan Cunningham, the New South Wales Colonial Botanist who visited Paihia and met William Colenso in 1838. His letter — entitled 'Rough notes collected from the New Zealanders (by the aid of missionaries), on the habits of the *Apteryx Australis*, a bird of New Zealand, closely allied to the *Struthionidae* and named by the native inhabitants *Kiwi*' — was read out, providing a full description of a 'most remarkable bird'. It inhabited the 'densest and darkest forests', reposing during the day either beneath long sedgy grass or hidden in hollows at the base of trees. The bird's solitary egg was about the size of a duck's, and its cry similar to 'the whistling made by boys with the help of fingers placed in the mouth'. Cunningham also described how the Maori captured the kiwi during their 'frequent night-prowlings in the woods' and how as a result it had been 'extirpated' in some districts where it 'once abounded'. If Owen was present that evening he may have been intrigued by Cunningham's concluding observations: 'Some natives of the country at East Cape, on the Coast, south of the

Bay of Islands, who are residing with the Church missionaries at Paihia, on its southern shore, observed that the Kiwies of their forests are much larger and more powerful birds than my specimen taken on the Hokianga river.' This led Cunningham to wonder: 'Might not those southern birds be of a distinct species?'[18] He would never know the answer to this question, because he died in that same year, and was buried in Sydney. Cunningham's younger brother Richard had preceded him as Colonial Botanist, but he died under unusual circumstances whilst visiting the Bay of Islands in 1835. He was said to have 'a singular facility for losing himself in the bush when intent on botany'.[19]

Richard Owen was well aware of the unusual birdlife in New Zealand when John Rule called on him in October 1839. But the identification of the first moa bone hardly got off to an auspicious start, with Owen being bothered both by the other's sudden appearance and claim that his specimen was from some eagle-like bird of flight. This would have necessitated a bird of truly monstrous proportions, and the Professor suspected (correctly) that such monsters might only be found in Maori mythology. Having listened to these far-fetched suggestions, it is not surprising that Owen did not appear to inquire into Rule's background. As mentioned, for some reason John Rule did not feel the need to advise Owen that he too was a qualified surgeon, included in the Royal College's regularly published lists of Members since 1806.[20] Perhaps it is unexceptional then that Owen would later insist on referring to Rule as 'the vendor', and give him little credit for suggesting that the bone was from a bird – albeit a flying one of the 'eagle kind'.

Thus John Rule's bone had a hesitant introduction to European science, but its timing was exquisite. Had it reached London any earlier it would have doubtless impressed all those Rule showed it to even less than it did in 1839. By then, at least the appearance of the kiwi might have alerted them to the possibility of other unusual birds in New Zealand, although nothing of the order of the moa. The existence of the smaller bird was not yet a valid basis for suggesting that much larger flightless birds – or their remains – might also be found there. As Owen observed later, 'this [kiwi] bird was barely the size of a Pheasant; and "the bone" indicated a bird as big as an Ostrich.'[21]

The irony is that there was plenty of evidence of the moa there for the finding, as would shortly be apparent. It is surprising that such bones escaped European attention – apart from Joel Polack's – for so long, and that the first to be collected was a less than impressive specimen. But had it reached London a few years later than it did, it is likely that it would have been overshadowed by another of Owen's taxonomic triumphs, the identification of an order of even larger and more remarkable creatures.

These first came to light in 1824 when the Rev. William Buckland named a recently discovered ancient terrestrial reptile – the carnivorous *Megalosaurus*. The following year, surgeon and amateur naturalist Gideon Mantell named a second genus, the herbivorous *Iguanodon*, and in 1833 came up with a third, *Hylaeosaurus*, also a herbivore. This impressive trio were lumped in with lizards until 1842 when, on account of their large size and the fact that each had an unusually strong pelvis ideally suited for terrestrial life, Richard Owen concluded that they represented a distinct 'tribe or sub-order of Saurian Reptiles' which he named *Dinosauria*. This extrapolation, based on slender evidence, was seen as further proof of Owen's exceptional deductive powers. He believed his beasts were a remnant from an age when many such creatures ruled the earth, and he saw them as being powerful and well-designed. The name 'dinosaur' was therefore borrowed from the Greek *deinos* (terrible) to suggest that these lizards were 'fearfully great'. It seemed an entirely appropriate choice, although Stephen Jay Gould suggests Owen had an ulterior motive and was cunningly enlisting these dinosaurs to support him in a much bigger philosophical debate.[22]

As well as being awesome and majestic, Owen saw 'his' dinosaurs as intricate and well-adapted. They were therefore ideal ammunition for combating 'transmutation', the belief that species were continually changing into other species, in an early version of a theory of evolution. Owen clung to a creationist style of progressionism favoured by natural theologians, who credited God with the creation of more complex organisms for each new geological age. The Hunterian Professor believed dinosaurs were on his side.

By 'inventing' this new order of animals Owen had also appropriated and made his own a family that more rightfully belonged to pioneering geologist Gideon Mantell. But they would forever be associated with Owen, and they came to dominate the natural history museum – which was yet another of his personal triumphs. The name 'dinosaur' implied 'truly fearsome', but the etymologically agile Owen had another shade of meaning of the Greek *deinos* in mind two years later when he announced the name *Dinornis* for a new genus from New Zealand, interpreting it as 'surprising bird'.[23]

While Owen was busy naming his dinosaurs, an important link in the identification of the moa was about to be made. William Charles Cotton, who had been educated at Eton and Christ Church, Oxford, had decided to travel to New Zealand as chaplain to that country's first Anglican Bishop, George Augustus Selwyn. The pair sailed on Boxing Day 1841, and also on board was Mrs Martin, wife of New Zealand's first Chief Justice, Judge William Martin. He had left for the colony eight months earlier, along with William Swainson, who was taking up his appointment as that country's second attorney general. Both Martin and Swainson were known to Owen, who referred to them variously as his 'valued' and 'most excellent and esteemed friends'.[24] He may have first met them at Lincoln's Inn Fields, although Swainson was also a fellow Lancastrian, five years Owen's junior. Martin was also a close friend of Selwyn, which may explain why he chose to go to New Zealand, and Cotton was also considered a 'friend' by Owen.[25]

Nearly six months after leaving England, Selwyn and Cotton arrived in Auckland. They soon sailed to Paihia, where they met missionary Henry Williams, and three weeks later Cotton was at the Poverty Bay home of Henry's younger brother, William. The pair discussed the New Zealand bird, 'of which [Owen] described a single bone',[26] and Williams showed his colleague the basket of specimens in the next room which were awaiting shipment to Oxford. On 11 July 1842, when back at Waimate, in the Bay of Islands, Cotton wrote to Owen telling him what he'd seen and advising that William Buckland could anticipate a certain osseous consignment.

If Richard Owen made a point of attending Zoological Society meetings when he had something important to announce, such was certainly the case on 10 January 1843. As Vice President and in the chair, he read a portion of the letter he had recently received from William Cotton. If he had harboured any doubts about the existence of such birds in New Zealand, then this letter brought the news he needed. Cotton described the hoard he had seen at William Williams' house, but did point out that 'no bones of the wings' had been found. On the basis of what he saw Cotton did not believe the moa was extinct, and ventured that he would not be surprised if the Zoological Society were to send out an army to 'take the monster alive'.[27]

With advance warning from Cotton, the first crate of moa bones from William Williams was eagerly awaited in England. At Buckland's invitation, Owen and William Broderip – a barrister and enthusiastic natural history collector who had an unrivalled cabinet of treasures at his chambers in Gray's Inn – were present at the grand opening. And such was their excitement that they hadn't finished the job when it was time for dinner, moving Broderip to quip, 'we supped upon them'.[28] Finally, after a wait of some three years, here was all the proof that was needed. With the later safe arrival of the second crate, there were now 47 bones, which the excited scientists compared with those of the ostrich, emu, rhea and apteryx, and concluded that they represented a new genus. Owen already had a name in mind, and set to preparing a new series of scientific papers. One of these was communicated at a meeting of the Zoological Society on 28 November 1843, and when published the following year included 13 plates, some necessarily in gatefold format to present the larger bone specimens at natural size.[29] A start could now also be made on the bird itself, and Owen managed to put together the legs of *Dinornis giganteus*. This assemblage naturally incorporated the largest single bone sent – the 2 feet 10 inch tibia – and suggested a complete bird at least 12 feet high. Nearly 90 years later it was described as 'one of the greatest curiosities of the British Museum'.[30]

At the Zoological Society meeting of 28 November, Owen went over the discovery of the *Dinornis*, recapping his first paper on the subject, read nearly four years earlier. On this occasion he made no reference to John Rule, and his paper began as had that first one: 'The fragment is the shaft of a femur, with both extremities broken off.' Then followed the letter from Cotton, to which Owen added another of 28 February 1842 written by William Williams to accompany the bone consignment to William Buckland. The Rev. Williams gave his story on the bones, making no reference whatsoever to William Colenso's involvement. According to Williams, about three years earlier when he was on the East Coast, he heard about an 'extraordinary monster' which was said to occupy an 'inaccessible cavern' near the Wairoa River and went by the name of 'Moa'. He regarded the story as a 'fable', until two bones were brought to his attention. Williams gave 'a good payment' for the second of these, which induced local Maori to 'turn up the mud at the banks and in the bed' of a local stream and uncover a large number of bones of various sizes. Williams compared them with the bones of a fowl, and 'immediately perceived that they belonged to a bird of gigantic size'.[31]

In his letter to Buckland, William Williams also estimated that the greatest height attained by the bird was 'probably not less than fourteen or sixteen feet', while the enclosed leg bones indicated a height of 'six feet from the root of the tail'. He concluded his letter with some 'worthy' information that had recently come to his attention. An American had told him that the bird still existed near Cloudy Bay, at the north-eastern corner of the South Island, the site of sealing and whaling operations since 1826. The same informant had also heard that the natives there had told an English whaler about a bird of extraordinary size which was only seen at night. Three men went to investigate and saw the creature, coincidentally estimating it to be 14 to 16 feet high. One of the men suggested going closer to shoot it, but the bird shortly took fright and 'strode away' up the side of a mountain. Not surprisingly, this account made an impression in London, and the Zoological Gardens reportedly entertained (high) hopes of obtaining a living specimen of the moa, the 'fourteen-foot high ostrich' from New Zealand. But unfortunately, 'London was never to feast its craving for the marvellous on such giant poultry'.[32]

If Owen had initially dismissed Rule's specimen as a beef bone, he was not alone. It was reported that 'more than one eminent naturalist of this metropolis' considered it to be no more than the marrow-bone of an ox.[33] But all such doubt was swept away with the arrival of the bones from William Williams. If the *Quarterly Review* is to be believed, all the anatomists' Christmases had come at once, and: 'that whitest day will ever be remembered. As each bone of the feathered giant was taken out it was impossible to repress exclamations; but when the enormous tibia came within our grasp, it was flourished aloft with a shout of wonder and joy that made the Museum ring again. Fortunately, we wore no wig, as dear Mr Oldbuck [a character in Walter Scott's *The Antiquary*] did, or it certainly would have been hurled upwards, where it would have ornamented one of the many antlers which overhung us'. After order had been restored, Owen selected the largest tibia to hold when he sat for a portrait painting commissioned by Sir Robert Peel for his 'Gallery of Modern Worthies' at his home at Drayton Manor.[34] Prime Minister from 1834–45 and 1841–46, Peel had also been able to keep an eye on Owen's progress as Trustee of the Hunterian Collection from 1824–51.[35]

Richard Owen's original deduction, which had initially failed to convince certain of his colleagues at the Zoological Society, was now the subject of unstinting praise. The 'colossal remains' of this 'feathered Goliath' could now look forward to taking their place alongside the 'Irish Giant' in the Hunterian Museum's pantheon of skeletons. According to one admirer: 'It is curious and instructive, with these wonderful bones before one, to look back to Professor Owen's description of the fragment of bone which first came under his notice, and to read the deductions which he drew from it. Entirely in the dark, with the exception of the glimmering light which he extracted from that fragment (the mere shaft of the bone, be it remembered), every word that he then wrote has come true to the letter. Long ago he showed us the outline, which he had drawn, of what the ends of this fragment of a femur ought to be; and it is but just to this acute and deep-thinking physiologist to say that if the drawing had been made from the perfect bone it could hardly have been more accurate.'[36]

Another who was present at that glorious moment of confirmation put it more succinctly: '[Owen] took, in our presence, a piece of paper, and drew the outline of what he conceived to be the complete bone. The fragment, from which alone he deduced his conclusions, was six inches in length and five and a half in its smallest circumference; both extremities had been broken off. When a perfect bone arrived, and was laid on the

paper, it fitted the outline *exactly*'[37] And in a letter to Sir John
Franklin, dated 25 February 1843, William Buckland offered his
version of the opening of the first crate. The freshness of the bones
indicated that they had only been languishing for a short time in
the mud from which they were extracted, and so they gave 'strong
hopes' that the living bird might yet be seen 'striding among the
emus and ostriches in Regent's Park'. As for the famous first femur
fragment, the missing terminations which Owen had drawn now
exactly matched those of a perfect bone. Such mastery of nature
was achieved, 'not from any guess, but from severe philosophical
induction'.

By coincidence, extracts from Buckland's letter were published
in the same 1846 issue of the *Tasmanian Journal of Natural Science* as
William Colenso's much delayed paper, 'An Account of some
enormous Fossil Bones . . .'[38] In his paper as completed on 1 May
1842, Colenso suggested the moa might provide the missing link
between the rhea, emu, cassowary and kiwi.[39] He could now read
Buckland's confirmation that the kiwi (*Apteryx*) was the moa's
'closest existing relation', and that Richard Owen had named the
bigger bird *Deinornis* [*sic*].[40]

The kiwi now played a part in explaining its larger relative.
Speaking to the Royal Institution of London on 'The Wingless
Birds of New Zealand', Richard Owen explained that while he
had now received various leg and foot bones of the ostrich-sized
struthious bird, no wing bones had come to hand. He therefore
concluded that this bird was a gigantic version of the *Apteryx* of
Australia [*sic*]. He referred to a kiwi in the collection of the
Zoological Society, suggesting its long beak resembled the bill of
a woodcock, its legs were like those of a fowl, and its trunk like a
cassowary's. And for those who doubted the existence of the dodo
on anatomical grounds, Owen offered the moa, which had the
same sort of features, except its beak resembled that of a vulture.
Owen illustrated his talk with a 'conjectural diagram' of the moa,
supposing its height to be 14 or 15 feet from head to foot. He also

spiced up his talk with the suggestion that the bird had been eliminated by New Zealand's first settlers. Having acquired a taste for animal food but having none at hand, they then 'took to eating one another'.[41]

The Revs William Cotton and William Buckland had suggested the moa might not be extinct, and in 1845 a similar suggestion was made by Professor Edward Hitchcock of Massachusetts. His speciality was ichnology, the study of fossil tracks and traces, and he went on to amass the largest dinosaur footprint collection in the world. Hitchcock was now of the opinion that the enormous birds' nests – made of sticks and measuring upwards of 26 feet in circumference – found by Captains Cook and Flinders on the coast of Australia were the work of moa. Nothing if not original, this idea was reprinted in the *Tasmanian Journal of Natural Science*, but quickly discredited when the nests were explained as the work of the sea or fish eagle, which unlike the moa,[42] was known throughout coastal Australia.

Richard Owen enjoyed the accolades that followed confirmation of his identification of the moa, but the subject still caused him a few problems. Speaking to the Anniversary Meeting of the Ipswich Museum in 1850, he began a lecture 'On the Gigantic Birds of New Zealand, and on the Geographical Distribution of Animals' with the now standard recap of his famous discovery. As usual there was no mention of John Rule, but William Williams and Walter Mantell came in for commendation for sending specimens which had been received in 'unexpected abundance and perfection'. Owen showed specimens and magnified diagrams, and then moved to the main point. He suggested that if all terrestrial animals had diverged from one common centre 'within the limited period of a few thousand years' it would be reasonable to expect their current location to reflect their respective 'powers of locomotion'. If so, flightless birds might have moved the least distance from the original centre of dispersion – which Owen located 'somewhere in the south-western mountain range of Asia' – and might also still be associated with one another. But this was certainly not the case, for the flightless birds had travelled furthest of all, to the very extremities of the earth. How, Owen wondered, could the cassowary, without the benefit of webbed feet, have crossed the hundreds of miles of ocean between the continent of Asia and the islands of Java and New Guinea? And how did the dodo manage to get to Mauritius, and the emu to Australia? Similarly, the rhea would have had to traverse the inhospitable expanse of Siberia to reach the northern part of America, and from there travel overland all the way to South America. Adding to the mystery, none of the original migrating rhea stock had decided to stop and settle on the prairies or any other part of North America during their long walk south.

Saving the hardest part for last, Owen turned to New Zealand, 'the abode of the little Apteryx'. How could such a bird – with wings reduced to the minutest rudiments, webless feet and feeble swimming abilities – have ever crossed the stretches of sea that lay between Asia and New Zealand? Now running out of time, Owen attempted to explain this dilemma with reference to terrestrial quadrupeds. Dividing the dry land of the planet into six areas, he pointed out that New Zealand was remarkable for its total lack of any such aboriginal species. Whereas Australia was rich in fossil remains of marsupial animals, New Zealand had none. However, it had no shortage of remains of wingless birds, some of which towered to 'the extraordinary height of eleven feet' – which suggested he was not convinced by William Williams' more generous estimations. The audience no doubt gasped at such a prospect, and went home happy, even if Owen had been unable to explain how such big birds had got to New Zealand in the first place. Such challenging concepts as Gondwana, continental drift and the origin and evolution of species would have to wait for another day.[43]

# 7

## THE LONE FRAGMENT
### The Famous Femur Spends Time in the Country

*I am sorry that I can find out nothing positive about the 'bone' Professor Owen wrote about. My Grandfather's and Uncle's collection of fossils and minerals has been packed up for many years and my cousin who is now abroad writes word that he does not remember anything about it but in the summer he hopes to look over the Collection and will remember it.*

RICHARD BRIGHT, 1870[1]

Following Napoleon's defeat at Waterloo in 1815, victorious Britain began sliding into an economic depression. For those who could see no way out of the gloom, enterprising businessmen promoted the idea of emigration to less populated parts of the world. Such a group approached Dr John Rule in St Day, Cornwall, seeking his interest in travelling as a surgeon on a ship carrying 234 female emigrants to Australia. In lieu of payment he was offered the prospect of land in or near Sydney, and employment in the public service of New South Wales. Rule took it on, and he and his wife and three adult children sailed on the *Layton*, arriving at Port Jackson in December 1833. Sydney then had a population of some 16,230, and its thoroughfares, which ranged from dust-bowls in the summer to bogs in the winter, were also distinguished by their traffic of bullock wagons and convicts marching to their enforced labours. Anticipating his new start down-under, Rule approached the Governor of the colony regarding the land and employment he believed had been promised to him, but learned there was neither any record nor even likelihood of such an arrangement. He set himself up in a medical practice, but by 1839 had decided to return to England – with a parcel of curiosities in his luggage. If he had enjoyed a free passage out to Australia, he would have had to pay for the trip back. Now desperately short of funds, Rule planned to capitalise on his bone.[2]

After its presentation to the scientific world on 12 November 1839, John Rule's 'unpromising fragment' might now anticipate further developments. However, the Royal College of Surgeons had already declined to purchase it for their collection and Owen himself passed up the opportunity to secure the object of what would be one of his greatest anatomical triumphs, pleading a personal shortage of funds: 'it was not convenient to me, in 1839, to pay the sum out of my own pocket.'[3] It is unsurprising that Owen did not feel inclined to spend the equivalent of some one and a half weeks' income on a small bone,[4] especially when he could hardly do so without the knowledge of his father-in-law, Hunterian Conservator William Clift. There was also a professional reason, for when taking up his duties at the Museum Owen had determined not to form a private collection.[5] But he now promised Rule that he would commend his specimen to others, through the good connections of his friend and colleague William Broderip.[6] Owen now waited for 'confirmatory' bone material from New Zealand, while Rule hoped for a quick sale.

Geology was in its infancy in the early 19th century, with several schools of thought on the origins of rocks. In particular there were the colourfully named Neptunists, Vulcanists and Plutonists, who put it all down to water deposits, volcanoes and magma respectively. Systematic observation became the key, as promoted by such pioneers as Gideon Mantell and William Buckland, and the science was firmly established with the publication of Sir Charles Lyell's *Principles of Geology* in 1830–33. He insisted that the present provided a key to the past, and did away with the earlier need for dramatic catastrophes to explain changes to the earth. Instead, Lyell maintained that the earth was much older than had been believed and changed its form through such interminable processes as erosion. Lyell's ideas proved extremely influential on the young naturalist Charles Darwin, who took the first volume of this landmark book with him on the *Beagle*, and had later parts shipped to him during the course of the voyage.

As well as a new science, geology was also a fashionable hobby. Cabinets and drawers were filled with tasteful arrangements of acquisitions, either personally accumulated on Grande Tours and rural rambles, or obtained through obliging dealers. As leisure pursuits grew, so did this interest in the earth beneath. Fossils and minerals were sought for reasons of rarity and beauty, as well as evidence of God's handiwork and His crafting of the earth to make it a suitable place for mankind. Many such specimens were to be found in the collection of the Bright family of Bristol, now identified by William Broderip as a likely purchaser of John Rule's bone. But if there was an early expression of interest, no quick sale ensued. On 31 March 1841 Rule wrote to Robert Bright at Weston Super Mare, giving him a full background to the specimen on offer.[7]

It seems John Rule had already shown his specimen to Robert Bright, and now began his sales pitch with some background to his fragment, this 'thigh bone of a Bird, now long extinct'. It was known as 'A Movie' by the natives of the country 'Ikana Movie', which translated as 'land, country, or island, of the "movie"'. The first settlers of that country – now known as New Zealand – had come from elsewhere, either as 'emigrants, or fugitives, or felons, or castaway fishermen'. But whatever their origins, Rule had great faith in their traditions, which were 'free from the embellishments of fiction'. Although 'destitute of scriptory aid' they had carefully preserved their 'tales of former times', and in Rule's opinion there was 'ground for reliance on the truth of their traditions'.

Rule explained that the 'movie' was 'a bird of the eagle kind', and also provided Bright with information he may not have shared with Richard Owen. This was not a migratory bird, bird of passage or an aquatic bird, and at a certain time of the year it 'took wing to the mountain forests'. He then came to the point: his 'unique fragment' was 'not equalled by any other in the world', and claimed its value had been variously put at between 10 and 50 guineas. He now offered it for 'the small sum' of £7, and took the opportunity of mentioning that he also had two crab-backed spiders from Norfolk Island which he was willing to part with for ten shillings, even though they were 'worth one pound'. To support the value of his bone, Rule then took the remarkable step of providing the potential purchaser with the full 800-word text of Owen's first paper published on the bone, in the *Proceedings of the Zoological Society* in 1840[8] – entirely written out by hand. We have to assume that Owen had not bothered to send Rule any complimentary copies of this landmark article, perhaps because he did not think the 'bringer of the bone' would have any further interest in such matters. But even if Owen had wished to extend this courtesy, he may have had trouble finding Rule's address at this time.

Having provided Bright with the full scientific context, Rule then offered his own views on the matter. For a start, he considered Owen's comment that the 'movie' was a sluggish bird and heavier than the ostrich a 'gratuitous assertion'. Neither were its bones found in the 'banks of rivers', and nor was the bone in question a fossil. He also took issue with Owen's inference that the femur of the ostrich was equal in size to that of the movie, for when he and Owen compared his fragment with the largest bones of the Hunterian Museum's ostrich they found the movie's had the larger circumference. And while Rule had entrusted his 'valuable piece of natural history' to Owen's care so he could show it to the Council of the Royal College of Surgeons, he now claimed he had been unaware of the Professor's plan to both exhibit it to the Zoological Society and submit a paper on the subject. Had Rule known of these plans he would have insisted that he be allowed to accompany Owen to the meeting and offer all his information on the subject. This, he suggested, would have rendered Owen's account 'more correct, and consequently, more interesting to naturalists'.

Six weeks later, Rule had not heard from Robert Bright and so wrote again, twice, hoping there might still be some interest in his bone. He explained that he was shortly leaving the district and begged Bright to advise how much he was prepared to pay for the specimen. In his own words, Rule was 'greatly in need of money',[9] but good news was imminent. Another member of the family, Benjamin Heywood Bright of Ham Green, near Bristol, now made an offer. Rule accepted and asked where he should deliver his 'unique specimen in natural history'. Again he referred to what he hoped was the temporarily 'exhausted state of [his] circumstances' and offered to 'shew testimonials of moral and professional character'.[10]

In his letter of acceptance to Benjamin Bright, Rule mentioned that he also had for sale a 'war-club of stone' and a 'curiously carved box', two of the Maori artefacts John Harris had given him in Sydney over four years earlier. While the subsequent fate of these items had little bearing on that of the bone, the 'war-club of stone' had the rare claim of being of both ethnological and ornithological importance. Identified by Owen as 'a jadestone weapon peculiar to New-Zealanders',[11] it was this item that suggested to him that Rule's bone might have a connection with that same country after all. There was now a strong element of doubt that the item might not just be a piece of beef or marrow bone as Owen had first insisted. He now paused to reconsider it, instead of dashing off to deliver his lecture, and most likely leaving the scientific discovery of the moa to another time, place, and perhaps another anatomist. The influential club, now recognised as a greenstone mere, was likely to have been a valuable item in its own right, and we might therefore wonder what became of it.

While the British Museum had turned away John Rule's bone, it would surely have been more interested in his four Maori artefacts. Since the early 1770s that institution had benefited from several donations by the Lords of the Admiralty of items brought back from newly discovered lands in the Pacific. On 7 June 1776 an assortment of 'natural and artificial curiosities' arrived via Captain James Cook and Charles Clarke, followed shortly by more of the same from Joseph Banks. With such generous donations, the British Museum may have had little need or inclination to pay for these items, as the cash-strapped John Rule would have certainly demanded. While there is no record of any Maori artefacts corresponding to John Harris's scant descriptions entering the British Museum at this time, they may have done so later via another route. In 1878, for example, the Museum received the collection of Sir Samuel Rush Meyrick, and although it was particularly rich in oriental armour, it did include several Maori items, among them a carved wooden box, with 'a rude head at each end', two wooden 'fiddle-shaped' mere, and two green basalt mere.[12]

Having finally arranged the sale of his bone, John Rule wrote an article for the *Polytechnic Journal*, identifying himself as 'J. Rule, late surgeon of free emigrants'.[13] Although entitled 'New Zealand', it was an excuse to tell the story of his bone fragment, which occupied nearly half the piece. He began by noting that, according to the charts, his nominal subject lay 13,340 nautical miles from London. But in actual fact the journey varied from 16,000 to 19,000 miles. He demonstrated the European's difficulty of transcribing an unfamiliar language when he gave the Maori name for New Zealand's South Island as 'Tavai Poennammoo'. It was in fact Te Wai Pounamu, so named because the valuable pounamu, or greenstone – or jadestone – was found on the West Coast of that island. The 'batoo powatoo' that Harris had passed on to Rule also reflected this, being a misinterpretation of patu pounamu, 'batoo' being a Maori club, which Harris translated as meaning 'to strike or kill'.

John Rule was obviously fascinated by nomenclature. He disapproved of the European custom of foisting familiar names on newly 'discovered' lands, much to the surprise of their 'native proprietors'. He noted the unimaginative – and, fortunately, short-lived – designation of New Zealand's three main islands as New Ulster, New Munster and New Leinster. Like Joel Polack before him, Rule had a few suggestions of his own when it came to names. For a start, he believed New Zealand would be better off as 'Salubria', derived from the Latin 'saluber', meaning healthful. He felt such a name was consistent with the physical condition of Europeans who now lived there alongside the native 'anthropophagi (man-eaters)'.

As he had for Robert Bright's benefit, Rule provided *Polytechnic Journal* readers with the complete text of Owen's first account of his bone, and also drew on the Professor's next paper, which included 'correct views' of the bone lithographed by George Scharf.[14] Rule acknowledged the 'great care and circumspection' of Owen's method, but naturally felt obliged to correct him on a few details. For a start, the full name of the bird in question was 'a Movie' – the prefix 'a' being important. And on the basis of 'the authority of a respectable person' who had resided and travelled in New Zealand for several years – obviously his nephew, John Harris – Rule believed the bird had long been extinct. He also believed his six-inch fragment to be the largest yet found, unaware that two crates of much larger specimens were being readied for their journey to Oxford, but Rule also corrected Owen's suggestion that his bone was a fossil, for it had not undergone any such change, and noted that it was he who had drawn Owen's attention to the distinctive texture of its interior. He now pointed out that this 'cellular lamella' continued throughout the cavity of the bone, which was not the case in the femur of the ostrich.

John Rule again took issue with Owen's deduction that the bone was a relic of a sluggish bird, like the dodo, and heavier than the ostrich, sticking to the natives' claim that it was a bird of flight. He felt it was neither a swimming or diving bird, but that the fragment was insufficient evidence upon which to conclude whether it was swift, slow or agile in its movements. He then noted that Owen had requested a loan of the bone to show it to the Council of the Royal College of Surgeons and – contradicting what he earlier told Robert Bright – that he gave his consent for Owen to exhibit it to the Zoological Society.

Rule's wide-ranging article also considered the possibility of the Asian origins of New Zealand's first inhabitants. If descended from Chinese, Chinese Tartars or Japanese, he surmised that their first landfall in New Zealand would have been on the present North Island, and on its 'east or Asiatic side', at or near the 'embouchure' of a river. As it happened, his bone had been discovered on the East Cape of the North Island. Whatever their origins, Rule was sure the language of the first New Zealanders had a common root with French. As evidence he offered 'mauve', the seagull in Norman French, while our own word 'move' obviously related to the motion of that bird, skimming through the air with 'no apparent flapping of its wings'. And although the 'movie' bird had long gone, Rule saw continuing proof of its former existence in everyday speech, as in a European's 'movie' wife. He was obviously confusing 'movie' with 'Maori', a word which certainly did enter the everyday language.

Shortly after the publication of his article in mid-1843, Rule read a report of a meeting of the Oxford-based Ashmolean Society at which William Buckland had announced the arrival of letters from New Zealand telling of the discovery of the bones of a large bird. Rule wrote to Buckland, asking if the latter could provide more information, identifying himself as the person who had provided the first fragment and offering to forward Buckland a copy of the latest issue of *Polytechnic Journal*.[15]

Little more is known of Rule's movements at this period, apart from the fact that on 3 August 1848 he called on Gideon Mantell. He asked to see his 'Moa's bones', but they had recently gone to the British Museum. Rule told Mantell that it was he who had 'brought home' the first fragment of moa bone, and had pointed out to Owen that it was from a bird. He also claimed that while Owen was offering the bone for sale to the Council of the College of Surgeons he described it without his permission. Mantell noted in his journal that Owen had told him that the College had bought the bone for £10, and now Rule was telling him that he had sold it to 'Mr Bright of Bristol' for £3! Such disclosures came as no surprise to Mantell, who had experienced similar treatment at the hands of the Professor. He lamented in his journal: 'This is worse and worse – what am I to believe? Is it possible Owen can have told me such audacious falsehoods, and which, in my ignorance I have promulgated.' In fact, on the same day as Rule had called, Mantell had received a letter from Owen requesting information on a new specimen in his (Mantell's) possession. Affronted, the latter reminded himself: 'I must avoid this man in future: but it is very sad thus to be compelled to become as reserved and selfish as the character I despise.'[16]

For some idea of John Rule's subsequent movements we are indebted to the much later researches of Thomas Lindsay Buick, who was born in Oamaru, New Zealand in 1865. From his involvement in journalism he began a series of regional histories, and also wrote two books relating to his lifelong interest in music. An opportunity of another kind beckoned in 1928 when Buick was asked to write an introduction for a reprint of a thriller written some nine years earlier by Wellington journalist

Ivan M. Levy. *Cornered by a Moa* was set in the wild interior of the North Island, and involved a confrontation with a 'belligerent' bird. The 'little romance' was 'printed and reprinted', and its popularity in America, South Africa and Australia was believed to be due to the fact that it at long last provided evidence of a live moa in New Zealand. There was a sequel, *Capturing a Moa*, but the public would not be fooled so easily by this one.[17]

Buick described *Cornered by a Moa* as 'one of the most successful hoax stories in our literature', and now pondered what he might contribute for the preface. The planned reprint never appeared however, but it had proved a fruitful exercise for Buick. He had realised that the moa represented a large gap in his knowledge of New Zealand, and undertook historical research to balance Levy's fanciful fiction. In so doing he discovered a new world, and 'looked over a high wall and peeped into a land of wonder and of mystery'. Inspired by the view, he produced a trilogy of books on the large extinct bird of New Zealand, the first – *The Mystery of the Moa: New Zealand's Avian Giant* – appearing in 1931.

Thomas Buick was one of a select group of New Zealand-born historians who in the early 20th century worked 'out of sense of duty and with little or no financial reward to make New Zealand's past readily accessible to the general reader'.[18] After writing *The Mystery of the Moa*, he realised there were some other unsolved enigmas relating to that bird, and chief among these was the identity of Dr John Rule. The few references then known to Buick were slight and unhelpful, some referring to Rule as a sailor or dismissing him as an 'illiterate seaman'.[19] Of course, the person who should have been able to provide all the information on Rule was Richard Owen, but as Buick noted, the Professor was 'singularly reticent and indefinite regarding his dealings with this gentleman'.[20] Buick began to make progress when he found an 1871 paper which made reference to the 'late' Mr Rule of Nelson, New Zealand, who 'took the first Moa bone to Professor Owen, and who had been represented in some quarters as an illiterate seaman ignorant of such matters, whereas he was an educated medical man, perfectly aware that the bone was that of a bird when he took it to England'.[21] Adding another twist to the story, Buick later discovered that the above reference to the 'late Mr Rule of Nelson' should have been been to 'Mr Rule, late of Nelson'.[22]

Working backwards, Buick next established that Rule had returned to Australia from England, and went from Melbourne to New Zealand in 1853. He landed in Wellington, before moving to the Nelson district, where he was certified as a Medical Practitioner in 1854.[23] A strong clue that this was the same John Rule was provided by a letter to the local newspaper. The writer, 'J. Rule, M.D., etc.', dismissed reports that settlers in Victoria, Australia, had seen the monster known as the Bunyip, claiming that he had already alerted the scientific world to another monster that was much less mythological, the colossal bird now known as the moa. As Rule had previously been in Melbourne – recognised as 'Bunyip country' – he should have known what he was talking about.[24] But in November 1857 the nomadic Rule had left Nelson for Sydney, and the trail went cold.[25]

Thomas Buick established that John Rule was born about 1779 in Cornwall, probably in the seaport of Penryn. Further, he came from a well-to-do family with strong nautical links, and was related to William Bligh, commander of HMS *Bounty* at the time of the mutiny and later appointed Governor of New South Wales. Rule continued this maritime connection by studying medicine and becoming assistant surgeon in the Naval Hospital in Jamaica. In 1800 he became ship's surgeon in the Post Office Packet Department, based at Falmouth, Cornwall; an appointment he held for nearly 25 years. He then practised as a surgeon – 'successively, if not successfully' – in several small Cornish towns until 1833, when he was attracted to Australia.[26]

Another source provides a different perspective on the

of its whereabouts.[33] Owen sent Bright a drawing – 'tracing'[34] – of the bone to stimulate the search, which now also involved other members of the family. Then in June 1872 Benjamin Bright reported that he did not think the bone was at Crawley after all, now believing it had gone missing before his father's death in 1843. He therefore advised Owen against going to Crawley in search of it, but now brought up the idea of offering his collection either to the British Museum or the Museum of Practical Geology in Jermyn Street, his only condition being that cartage to London would be met by the institution. The collection was housed in cabinets and drawers, which would be included, and Bright was confident of locating the keys for these. He believed the collection would be better off in Owen's hands, and asked for his thoughts on the matter.[35]

Already suffering from Benjamin Bright's lack of interest, the collection at Crawley was now under threat from another quarter. In 1870 the old house had been sold to Lord Ashburton, who began cannibalising it for its bricks and stones. According to Bright, it was only by the new owner's 'sufferance' that the 'fossils' remained at Crawley.[36] Owen responded, advising that the British Museum would be interested in Bright's offer. The latter wrote back on 1 July 1872 confirming this arrangement, but the bone still eluded him.[37] The following year an assistant from the Museum's Geology Department was sent to pack the collection in preparation for its move to Bloomsbury. He reported there were 50 cabinets and cases of (mainly) minerals and fossils, and others containing 'private papers, books and other objects, all more or less mildewed', the whole lot weighing about three tons. The assistant had the added challenge of getting the collection down 'two flights of awkward stairs'.[38] It seems the collection was still at Crawley in September 1873, when Benjamin Bright warned Owen that another winter there would 'seriously injure' it.[39] A short time after it was safely evacuated to London, Lord Ashburton sold the Crawley property, and the old house, once faced with stone and topped by 18 chimneys, was pulled down and replaced by a new structure in the Elizabetho-Jacobean style.[40]

At some stage the bone did turn up among the mildewed fossils and minerals rescued from Crawley. In October 1873 Richard Owen reported that he had received the permission of Benjamin Bright, Esq., to deposit in the British Museum this portion of 'bone of an unknown Struthious bird of large size, presumed to be extinct'. Owen provided the usual background to the specimen, but now there was some slight recognition of the (still unnamed) individual who had offered to sell it to the Royal College of Surgeons. The Professor acknowledged that his opinion that it was the shaft of the femur of a bird was assisted by 'the evidence' from the 'vendor' that made it at least probable that it had been found in New Zealand.[41]

With the bone in the collection of the British Museum, Richard Owen could now compare it with other moa bones, something he had been unable to do when he'd last seen it some 33 years earlier. At some stage he was photographed holding his historic specimen, alongside the reconstructed skeleton of a moa that had been discovered by gold-miners in Central Otago in the South Island of New Zealand in 1863 and presented to the York Museum the following year. We do not know if the famous femur ever went on display at Bloomsbury, but it was exhibited 'in a prominent position' in the new natural history museum in South Kensington until at least 1930, when it appeared to be 'in perfectly good condition'.[42]

# 8

# BOUND FOR THE ANTIPODES
## Emigrating to the Land of the Moa

*There is, probably, no part of the world which presents a more eligible
field for the exertion of British enterprise, or a more promising career of
usefulness to those who labour in the cause of human improvement,
than the islands of New Zealand.*

JOHN WARD, 1840[1]

On Sunday 12 May 1839 the 382-ton barque *Tory* nosed out
of Plymouth Sound. It had left Gravesend a week earlier,
answering cheering spectators on shore with a salute from its
eight guns. It was then towed by steamer to the mouth of the
Thames, and from there enjoyed a quick run of 38 hours to
Plymouth. This was only the *Tory*'s second voyage, and now under
the command of Captain Edward Chaffers, R N – who had been
master of HMS *Beagle* during its circumnavigation of 1831–36
– it would carry 35 crew and passengers to New Zealand. Among
them was the first trained scientist to live and work in that country, and one whose inquiries would inevitably lead to the moa.[2]

The departure of the *Tory* was the realisation of long-standing
settlement plans for New Zealand. Of course, that country was
hardly uninhabited, having received its first human settlers from

east Polynesia some 900 years earlier, but was seen as a suitable subject for a further round of colonisation, from Europe, after its rediscovery by Captain Cook. Such a scheme had been suggested as early as 1771 – by Benjamin Franklin – but nothing came of it. It was Britain's establishment of a penal settlement at New South Wales that revived interest in settlement on the other side of the Tasman Sea. After the end of the Napoleonic Wars, emigration was seen as a practical solution to Britain's growing problems of poverty and homelessness. The advancing Industrial Revolution had created much unemployment, with machinery now proving more economical than handwork, and displacing poorly-paid women and children. Landowners could now expect better financial returns by raising sheep, and consequently forced tenant farmers from their lands. Likely candidates for emigration were the new class of skilled but impoverished tradesmen, farmers and professional men. They saw New Zealand as their hope for a new life, while the wealthy back Home recognised it as a smart new investment.

Planning for organised settlement began with the first New Zealand Company, formed in London in 1825. Aspiring to what it termed 'sound principles of colonisation', it hoped the British Government would assist its aims by providing a military force. But Government had no such intentions, nor any other plans for New Zealand at that time, so the company had little choice but to go it alone. It took a more commercial approach and dispatched two ships to assess trading prospects. But this expedition achieved little, apart from recognising the prohibitive cost of exporting New Zealand timber to Britain, and casting doubts on the commercial value of the country's flax. It was a costly speculation, but members did not abandon the idea, attempting to retrieve their losses by a later revival of the company.[3]

The next significant development was the formation of the New Zealand Association, in 1837. This was the initiative of Edward Gibbon Wakefield, who had earlier spent three years in

Newgate Prison for abducting and marrying a 15-year-old schoolgirl. He turned his confinement to advantage by studying the subject of colonisation, with particular reference to New Zealand, believing it to be admirably suited for such a process. But his new association also had problems furthering its aims. It had a stern enemy in the Church Missionary Society, which argued that colonisation was always detrimental to the indigenous people, and would also undermine its own good work in New Zealand. A British Government offer of support was withdrawn when the Association refused to become a joint-stock company, and the matter was also subject to debate in a Select Committee of the House of Lords – at which Joel Polack gave evidence. The Association introduced a bill into Parliament, but it was soundly defeated. Undeterred, members of this and earlier thwarted enterprises were soon at it again, reforming as the New Zealand Colonization Company in August 1838.

By now there were nearly two thousand British subjects settled in New Zealand, and, according to John Ward, secretary of the latest company, a significant proportion were a 'worthless class of person', being runaway sailors, escaped convicts, keepers of grog-shops, desperadoes and 'other vagabonds of dissolute habits'. While his company obviously had a vested interest in highlighting the pitfalls of 'disorganised' settlement, Ward claimed the crimes committed in New Zealand by some captains of British vessels were 'so atrocious as to be hardly credible'. He acknowledged the good works of 'a few missionaries' and other well-disposed settlers, but the other lawless Englishmen only encouraged the natives' natural vices and taught them new ones. These undesirables also introduced diseases and gave the natives a taste for alcohol, and as far as Ward was concerned they deserved to be known as the 'Devil's missionaries'.[4]

Eventually the British Government was stirred into action, and decided to negotiate with Maori chiefs for the acquisition of sovereignty over New Zealand. This in turn forced the hand of what was now known as the (second) New Zealand Company, which needed to buy as much Maori land as it could and at the lowest possible price. Land was fundamental to Wakefield's plan, for the reselling of it would finance the emigration scheme. Anticipating the annexation of the country, the Company set up a committee to dispatch a preliminary expedition as soon as possible, and there was a flurry of activity as a suitable ship and crew were sourced.

The *Tory* would carry principal Company agent Lieutenant-Colonel William Wakefield, otherwise known as Colonel Wakefield, who had served a three-year imprisonment, for assisting his older brother Edward's abduction of a minor. Also on board was draughtsman Charles Heaphy, and Maori translator Ngatai, who had lived in England for the previous two years. In their final instructions to Wakefield, the Company directors expressed their concern for 'Naiti [*sic*], a New Zealand chieftain.' Though a 'complete savage when he arrived' in England, he had now acquired the manners, habits and tastes of a 'well-bred Englishman'. By treating him with respect, Wakefield would assist the 'continued relative superiority of . . . chief families', which would set an example for the 'lower orders'. Ngatai could therefore be conclusive evidence of the original New Zealanders' capacity for 'performing useful parts, and occupying respectable positions in a community of British emigrants.'[5]

Also on board the *Tory*, with instructions to report back with 'general information relating to navigation, geography, geology, botany and zoology, and the traditions, customs and character of the natives', was naturalist Johann Karl Ernst Dieffenbach.[6] He was born in Germany in 1811, and studied medicine at the University of Giessen. Apart from New Zealand Company business, he would have something else in common with the Wakefield brothers. He had also been imprisoned in Zurich for two months as a result of his involvement in student agitation for reform. He later qualified as a doctor of medicine and made his way to London, moving in scientific circles and getting to know Charles Lyell, Richard Owen and Charles Darwin. It may have been the latter who encouraged him to apply for the position of naturalist with the New Zealand Company, for which he was also nominated by two officials of the Royal Geographic Society.[7]

Farewell formalities for the *Tory* included a lavish tavern luncheon for 130 at the West India Docks. Diners drank the health of the young Queen Victoria, and speakers reminded them how British settlement would both enhance the glory of Her Majesty's reign, and bring the benefits of civilisation to an otherwise savage land.[8] To lead the way on this bold new venture, the three-masted vessel had a figurehead of the Duke of Wellington, hero of Waterloo. At sea, Captain Chaffers found the ship 'very stiff under canvass', but able to average eight knots with a good breeze. Some of that breeze would have been welcomed below decks, for the captain ordered the 'nitrous' fumigation of the ship to counteract the foul air, and he also noted that, fortunately, the crew washed their clothes regularly.[9] William Wakefield had hoped the journey could be achieved within 100 days, and in fact it only took 96. On 20 September 1839 the *Tory* arrived at Port Nicholson (Wellington Harbour), and it was now his responsibility to negotiate with local Maori for the purchase of land.

Three days before reaching her final destination, the *Tory* had reached Queen Charlotte Sound, at the northern end of the South Island. It was here that Ernst Dieffenbach applied the latest scientific methodology to this new country, when he ascended a hill and calculated its height − 1544 feet − by means of the temperature of the atmosphere.[10] When the *Tory* set sail to survey for another of the Company's planned settlements, New Plymouth, he got his first glimpse of that region's dominating mountain, Taranaki, named Egmont by Cook in honour of the First Lord of the Admiralty. It was from this expedition that

Company draughtsman Charles Heaphy painted his now well-known image of that symmetrical snow-clad cone flanked by tree ferns, *Mount Egmont Viewed from the Southward*.[11] The peak was a 'never failing object of attraction' for Dieffenbach and he was determined to climb it, against the advice of local Maori. They claimed it was 'tapu' and inhabited by ngarara (large reptiles) – which would surely eat him – and a mysterious bird known as the 'moa'. Undeterred by such creatures, Dieffenbach set out, but upon reaching the snow-line his two Maori guides refused to go any further. They squatted down and began to pray, for according to Dieffenbach none of them had ever before been so high. Another good reason for their reluctance to continue was the intense cold, for they had 'uncovered feet'. He pressed on without them, and became the first European to reach the top, on Christmas Day 1839. He took in the view from an altitude of 2521 m, and saw nothing untoward. Back on the ground he later wrote: 'The mountains are peopled with mysterious and misshapen animals; the black points, which the [Maori] sees from afar in the dazzling snow are fierce and monstrous birds; a supernatural spirit breathes on him in the evening breeze, or is heard in the rolling of a loose stone.'[12]

After his contract with the New Zealand Company had expired, Dieffenbach sought permission to continue working in the country, but his request was denied. In October 1841 he returned to England, where his two-volume book *Travels in New Zealand* appeared two year later. This landmark publication was the first general scientific account of New Zealand, and also reflected the author's concern for the welfare of the Maori people. It had the added distinction of being the first publication to refer to that country's extinct struthious birds as 'moa'.

In his book Dieffenbach included a realistic illustration of the 'Kiwi Kiwi', or 'Apterix [sic] Australis', in contrast to the earlier and standard penguin-like pose, and suggested it was a member of the same order as the extinct moa, or movie. His book was probably also first to acknowledge John Rule's part in taking the first evidence – 'a bone very little fossilized' – to London, and suggested it was Mr Gray at the British Museum who then sent the specimen to Professor Owen. Dieffenbach also claimed to have drawings of similar bones, and possibly of a claw, from the collection of his 'excellent friend', the Rev. Richard Taylor at Waimate. Such bones were found on the East Coast of New Zealand, and were brought down by rivulets from a neighbouring mountain called 'Hikorangi'. And whilst in the Rotorua district, Dieffenbach asked Maori people about local natural history and was told a 'curious tradition' whereby the last moa had been killed near a particular totara tree. His systematic approach extended to the inclusion of a Maori dictionary in his book, and among its some 2500 entries was 'Moa', defined as 'fossil bones of a struthious bird of that name'.[13]

A week before the *Tory* arrived at Port Nicholson, hopeful colonists had already left England. They went on a flotilla of five ships – the *Oriental, Aurora, Adelaide, Duke of Roxburgh* and *Bengal Merchant* – and at the last minute a sixth, the *Glenbervie*, was hired just to carry the freight. In all there were some 811 passengers, travelling either in barrack-like steerage accommodation or in first- and second-class cabins. They were about to undertake the longest emigrant journey in the world, and, notwithstanding the large Scots contingent, it was suggested that: 'None but Englishmen would have ventured on such an enterprise, and Englishmen only have the habits which would ensure its success'.[14] The Company Secretary felt it 'no exaggeration to assert' that for 'intelligence and energy of mind, as well as for rank and character in society', this body of emigrants had not been equalled since the days of the early colonisation of North America.[15] Indeed, the nucleus of this new colony would comprise persons 'of considerable property, and members of some of the oldest and most respectable families in the kingdom'.[16]

At a meeting in Glasgow in October 1839, intending emigrants were told they were marking 'one of the most interesting events in the history of our country – the colonization of a new and highly important island in the Southern Hemisphere, and the spread of the British race'. They were given a florid description of their destination, including the information that it had 'torrents affording an inexhaustible supply of water-power for machinery'. As for climate, a vast interior mountain range and an encircling ocean sheltered this favoured country from 'the shivering blasts of winter and the scorching heats of summer'.[17]

Another good reason for emigration was Britain's 'numerous and redundant population' of 'five-and-twenty millions . . . multiplying at the rate of a thousand souls a day'. One speaker foresaw the great day when colonies – those 'boundless regions' – would be peopled with 'the bone of our bone and the flesh of our flesh'. But he had an ulterior motive, for these 'true descendants of the Anglo-Saxon race' would be so imbued with British tastes that they would be forever dependent on manufactures which only Britain could provide.[18] In a sense this proved to be the case, for New Zealand maintained a heavy dependence on Mother Britain, both as a market for exports and source for imports, until the latter joined up with the European Economic Community in 1972.

There was no shortage of dramatic incidents on the way to New Zealand. *The Duke of Roxburgh* experienced six deaths among the emigrants, most of them infants. The *Bengal Merchant* also had a death, which was at least balanced by a birth and a marriage. Cramped conditions below decks were the cause of much quarrelling, with the *Adelaide* calling into the Cape of Good Hope so that four duels could be fought, while two antagonists on the *Aurora* agreed to postpone theirs until they reached New Zealand. All ships reached their destination safely, although a large dog belonging to the Hon. Henry Petre – the 19-year-old son of Lord Petre who had given evidence at the House of Lords Select Committee – was washed overboard from the *Oriental*. King

Neptune also claimed the captain of the *Duke of Roxburgh* in similar fashion. The *Aurora* enjoyed the quickest passage, 104 days, ten days faster than the *Oriental*, while last to arrive was the *Adelaide*, no doubt delayed by the demands of its duellists. When it finally entered Port Nicholson, the *Adelaide* was greeted by a southerly storm, giving passengers an early taste of that region's robust climate.[19] Later that same year they experienced another of its characteristics when the very foundations of the new settlement were shaken in a series of earthquakes. That event would have come as a shock to the new colonists, who had been advised that there was no reason to believe that the country was subject to such disturbances, there being 'no record of any within the memory of man'.[20] Later, those contemplating moving to New Plymouth were assured that: 'Vibrations, or slight tremblings of the earth, are occasionally experienced, but have never, in the memory of the oldest natives, caused the least damage.'[21] In fact, New Zealand has about 100 earthquakes a year, and in 1855 Wellington experienced the country's greatest tremor in historical times, with a magnitude of about 8 on the yet-to-be-invented Richter scale.

The *Oriental* arrived at Port Nicholson on 31 January 1840, and among its 154 passengers was Walter Baldock Mantell, a month short of his 20th birthday. Attracted by New Zealand Company propaganda, he saw emigration as an escape from an increasingly uncomfortable family life. His father, geologist Gideon Mantell, suffered chronically from ill health and the failure of his medical practice, on top of which his wife had recently deserted him. Walter trained as a surgeon in Chichester, and had sailed to New Zealand against his father's wishes. But perhaps all was not lost, for Gideon saw that country as a likely source of natural history specimens, which might prove to be to Walter's financial advantage, and at the same time advance his own scientific career back in England. While Walter was still halfway to New Zealand, Richard Owen presented a fragment of moa bone to the Zoological

Society, and before long Gideon would be suggesting that his son keep an eye out for further evidence of this large bird.

Walter Mantell arrived in Wellington with no particular career in mind, and worked variously as Postmaster and Clerk to the Bench of Magistrates. With such responsibilities he travelled to other New Zealand Company settlements at Wanganui and New Plymouth, at the same time developing his interest in New Zealand natural history. In April 1845 he wrote to his father complaining that he was 'penniless and without any prospect of employment',[22] but he would shortly be appointed superintendent of military roads at Porirua, some 21 km north of Wellington.

In December 1845 Walter Mantell announced to his father that he was 'going into the interior in search of the Moa!'[23] As for his chances, Gideon claimed that: 'If there is a live Moa, my son will catch it.'[24] And although unsuccessful in his search for a live specimen, Walter did secure large quantities of posthumous evidence. These he dutifully dispatched to his father, who described one consignment as: 'Perhaps the most extraordinary collection of fossil remains of struthious birds that has ever been transmitted to Europe.'[25]

In early 1847 Walter was at Waingongoro, on the South Taranaki coast, a site that had proved a rich source of moa bones for the Rev. Richard Taylor nearly four years earlier. As T. Lindsay Buick put it, Mantell 'did not scruple to appropriate some of the reverend gentleman's reserved collections'. But Mantell saw it differently, claiming that on his previous visit to this 'wide flat of undulating sand about 200 yards across' the surface was covered with 'bones of men, Moas, and seals, etc., which had been overhauled by the Rev. R. Taylor'. He now dug at the base of the cliffs, and came upon 'the regular bone deposit', but unfortunately, many perfect examples were so soft that they did not survive their excavation.[26] This frustrating experience may have recalled that of his elder sister Ellen when she accompanied their father on an expedition on the chalk hills near Brighton. After three hours of digging they exposed a huge bone, which measured 9 feet long and 30 inches in circumference, and was probably part of a plesiosaur or mosasaur. But when removed from the chalk cliffs, it disintegrated into a hundred pieces.[27]

At Waingongoro, Walter Mantell was also frustrated by the attentions of the local Maori, who trampled on the many fragile bones he had successfully excavated and laid out to dry. Such was the loss of specimens, he suggested they might be placed in the same category as 'the vandals who destroyed the Alexandrian Library'. The bones that survived the archaeological challenges at Waingongoro were packed up and carried overland to New Plymouth, where they were sorted and readied for their journey by sea to Wellington. Mantell did not accompany them on this voyage south, choosing to take the overland route in the hope of finding further specimens. It may have been on this occasion that he turned up a four-inch fragment of moa egg-shell, and suggested that a common hat would have served as an egg-cup for it: 'what a loss from the breakfast table!'[28] Back in Wellington and reunited with his specimens, Mantell prepared them for dispatch to his eagerly awaiting father.

Eight months later – in December 1847 – the Waingongoro collection arrived in London. Gideon Mantell appreciated that in order for the items to be of any value, financially or scientifically, they needed to be described by an authority. He had limited expertise in this area, so decided to make the magnanimous gesture of placing them at the disposal of the recognised expert – Richard Owen. Before long the Professor was able to announce that there was a lot more than moa in this consignment, including the skull of an unknown bird, which he named *Notornis* ('southern bird') *mantelli*. This bird was presumed to be extinct, but in 1849 Walter Mantell came across evidence to the contrary. During the period 1848–49 he was appointed to a government post dealing with the settlement of Maori land claims in the South Island, and his wide-ranging duties brought him into further contact with deposits of moa remains. At a coastal swamp at Waikouaiti, on the east coast some 42 km north of present-day Dunedin, he gathered a trove of some 500 bones. Among them was a pair of well-preserved tibiae with perfect toes attached, first sighted by whaler Thomas Chaseland. Of these treasures Mantell wrote: This unlucky Moa, happily for science, must have been mired in the swamp and, being unable to extricate himself, have perished on the spot.[29]

In the South Island Walter met a sealer with a collection of bird-skins, and recognised one as the *Notornis*. He learned that the bird in question was known as a takahe, and had been caught by a dog in Dusky Bay – now Sound – at the south-western corner of the South Island. Walter bought the unique skin – along with others of the kiwi and the soon to be extinct huia – and duly sent them on to his father.[30]

In 1851 Walter Mantell was able to obtain a second specimen of the takahe, shot by a party of whalers, and this ended up in a museum in Washington. Much later, in 1879 and 1898, two more birds were caught, bringing the known total to date to just four. No more was heard of the takahe, and by the 1930s it was presumed to be extinct. But that theory was overturned in 1948 when a colony of the birds was discovered in an isolated corner of the South Island. Takahe once lived and were hunted throughout New Zealand, but unlike the moa managed to defy extinction by retreating to remote mountain valleys. But even there these largest living members of the rail family are not safe, for introduced stoats and deer now eat their preferred diet of tussock grass.

By 1852 Walter Mantell had accumulated another collection of moa bones that rivalled that retrieved from Waingongoro. This hoard, which included at least 25 skulls of different types, was packed up in seven crates and dispatched in March 1853. But Gideon Mantell never got to see it, for he had died the previous November. The British Museum had no great need for further moa bones at this stage, for Richard Owen was still working his way through Walter's previous offerings from Waingongoro. In 1856, following the death of his younger brother Reginald from cholera in India at the age of 30, Walter took leave of absence from his positions as Commissioner of Crown Lands for Otago and Justice of the Peace, and went to England.[31] There he arranged for the British Museum to purchase the last collection of bones he had sent 'home', and also managed to meet Owen, who was about to be appointed Superintendent of the Natural History Department. With these bones, and the help of William Flower, the 'Articulator',[32] the Professor was able to reconstruct and place on exhibition the largest moa skeleton yet discovered, *Dinornis elephantopus*.[33]

In complete contrast was a small relic Walter Mantell found while gathering together his late father's remaining possessions in London. This was a tooth of the *Iguanodon*, the ancient terrestrial reptile, which Gideon's colleague Charles Lyell had taken to Georges Cuvier in Paris for identification in 1823. In 1859 Walter brought this historic fossil back to New Zealand, where it now resides as item MNZ GH 004839 in the collection of Te Papa Tongarewa Museum of New Zealand.[34] It could therefore be seen to balance another small and important item, No. 44639 in the Natural History Museum, South Kensington, which had made the reverse journey in the custody of John Rule some 20 years earlier.

Back in New Zealand, Walter Mantell entered politics, and for five months in 1861 was Minister of Native Affairs. He had been one of the most active and important links in the chain that furnished English scientists with evidence of the moa, a contribution that ended with the death of his father. But the person who benefited most from these valuable collections of bird remains from New Zealand was neither Walter nor Gideon Mantell, but the imperious Richard Owen. If Walter had enjoyed a cordial meeting with the Professor in London in 1856, he later

commented: 'He has made considerable blunderings and floun-
derings in his search after renown rather than truth.'[35]

For further insights into Walter Mantell's opinion on Richard
Owen, the former provided appended footnotes to his own copy
of the other's 1879 *Memoirs on the Extinct Wingless Birds of New Zealand*,
now in the collection of the Ornithology Department of the
Auckland War Memorial Museum. In the first sentence of the
preface to this volume, Owen pointed out – with the benefit of
hindsight – that the advantages of paying attention to any object
of natural history, 'however unattractive', were exemplified by a
fragment brought to him 40 years earlier by an 'individual', who
Mantell noted tartly was 'Mr Rule M.R.C.S.' Owen recounted how
his first paper on the subject was accepted for the *Transactions*, with
responsibility resting 'exclusively with the author', and published
in '1838'. This date was obviously wrong, for the bone was still in
Sydney at that stage, so Mantell corrected it to '9 40' – September
1840. Also, Owen's claim that 100 extra copies of the paper were
printed and 'distributed in every corner of the islands of New
Zealand' had long been an irritation to Mantell. In 1844, when
writing to his father on the subject of the moa, he regretted that
'no copy of Prof. Owen's paper on the Dinornis is in the country'.[36]
Now, some 35 years later, he responded to Owen's claim with
the comment: 'Impossible and untrue.' And giving credit to
another overlooked pioneer in the field, he added that even if
the Professor's paper had been available it 'contained no
information beyond or even equal to that previously known
through the island and in England since the publication of
Polack's book in 1838'.

# 9

## FOSSILS, FROGS AND GRAINS OF SAND
### Family Connections with the Moa

*Every grain of sand is an immensity − every leaf a world . . .*
JOHANN CASPAR LAVATER (1741–1801)[1]

The discovery of the moa involved two pioneering geologists who had a lot more than extinct birds in common. Both Gideon Mantell and William Buckland had family connections back to the land of the moa, which continued long after they'd departed the world of science. The pair were also colleagues of the formidable Richard Owen, although their personal relationships with the Professor were in sharp contrast. And even when their life's work was done, Mantell and Buckland maintained a most unlikely link with science, for selected parts from their worn-out bodies went on display in the Hunterian Museum.

As a single bone made its way from New Zealand to the British Museum in 1839, a large collection of specimens, the property of Gideon Mantell of Brighton, was being prepared for the same destination. Born a shoemaker's son in Lewes, Sussex in 1790, Mantell had developed a successful medical practice, specialising in midwifery. He was an extremely energetic doctor, managing to visit upwards of 60 patients a day and still find time for his main passion, the pursuit of natural history and geology, which he developed at an early age when gathering specimens from the Weald of Sussex. It is said that one day while Mantell was tending to a patient, his waiting wife Mary Ann noticed an unusual piece of rock lying by the roadside and showed it to her husband. He recognised it as a fossilised tooth, and traced its immediate source back to a local quarry, from where he later obtained further specimens. Mantell took his new find to a meeting of the Geological Society in 1822, convinced he was on the trail of an unknown animal.

In 1823 Charles Lyell took Mantell's mystery tooth to Paris to show the then undoubted authority on such things, Georges Cuvier. However, the great French anatomist pronounced it as no more than the incisor of a rhinoceros. Disappointed but not daunted, the determined Mantell investigated collections of specimens that might provide a clue, a search that led to the Hunterian Museum. Richard Owen would not join the staff there for another two years, and it would have been a most interesting encounter if it had been he who had dealt with Mantell. Instead, Owen's future father-in-law, Conservator William Clift, attended to the inquirer, but was unable to help. Fortunately, Mantell solved the first part of the mystery himself, for whilst at the Hunterian he noticed the skeleton of an iguana lizard with teeth similar to his own specimen. With this confirmation, a new beast was born. Mantell proposed naming it *Iguanosaurus*, but a colleague convinced him that *Iguanodon*, meaning 'iguana tooth', was a better choice. Mantell's account of the ancient beast was read before the Royal Society in 1825 – and later published in the *Philosophical Transactions* – and in that same year he was elected a Fellow, FRS, of the Society.[2]

Gideon Mantell moved his family the short distance from Lewes to fashionable Brighton on the south coast in 1833. He was anticipating further scientific recognition, which came two years later when his discovery of the *Iguanodon* and *Hylaeosaurus*

– which he'd chipped out of a great slab of quarried rock – was acknowledged by the Geological Society's highest award, the Wollaston Gold Medal. But if his career was on the ascendant, it would shortly be overtaken by that of Richard Owen, who was the 1838 recipient of the same medal – for his work in progress on identifying Darwin's fossils from South America.

A 'facile and prolific' writer,[3] Mantell published his first book, *The Fossils of the South Downs*, in 1822, with plates executed by his wife. In 1836 his *Thoughts on a Pebble, or A First Lesson in Geology* opened with an engraving of the author as Vice President of the Geological Society, posing proudly in front of his *Iguanodon* remains. The book included 27 exquisitely coloured 'lignographs', among them 'The Pebble', a humble stone 'as might be found lying in a lane'. Mantell mused on the past lives and wanderings of such fragments, seeing them as evidence of an earlier world. No natural object was too trifling, for it was from the contemplation of such that man might understand some of the 'grand truths' of the Great Designer. This was not an altogether new idea, for some 33 years earlier English visionary poet William Blake had written:

> To see a world in a grain of sand
> And a heaven in a wild flower
> Hold infinity in the palm of your hand
> And eternity in an hour.
> *Auguries of Innocence,* c. 1803

The contemplation of everyday objects, both large and small, had set Blake on the road to mysticism, while Gideon Mantell now believed an increased knowledge of creatures – from the mightiest to the most minute – would provide proof that nothing in the universe had been created in vain. These flint stones were 'medals of Creation', and each was a page from Nature's volume, waiting to be read. But these same stones could also be misinterpreted. Frustrated farmers and gardeners who regularly cleared their fields and plots of apparently fresh 'crops' of flint stones were inclined to believe that such bothersome objects 'grew' naturally. Stones rose to the surface in tilled soil, but did not 'grow', although Mantell explained that they could become encrusted and cemented together to form solid conglomerates. In his *Thoughts on a Pebble* Mantell also included a mysterious mollusc, referring to another scientist's 'admirable description of the recent Nautilus'. Interestingly, this was not the highly praised memoir produced by Richard Owen some four years earlier.[4]

As his commitment to science grew, Mantell neglected his medical practice. He soon became a doctor without patients and had little choice but to move his family out of its grand Brighton house, which he now planned to turn into a revenue-generating scientific institution and museum. When even that venture failed, Mantell faced the prospect of selling his 30,000 treasured specimens, the finest collection of its type in existence. Finding no likely takers he appealed to the British Museum, and with the support of such influential colleagues as William Buckland an arrangement was reached and the figure of £5000 agreed upon.[5] And so in late 1838 the Mantellian Museum closed and its contents were soon on the way to Bloomsbury. But Mantell's misfortunes continued to mount, for his wife had now deserted him, his eldest daughter had left home, and his son Walter, who had completed his training as a surgeon, was determined to emigrate.

Another of Mantell's books, *The Wonders of Geology*, appeared in 1838 and enjoyed six editions in its first ten years. It consisted of a series of lectures he'd given at Brighton, and perhaps its popularity was due to the fact that geology, whose language was 'more mysterious' than the other sciences, could now be seen to bear witness to the truth of the Bible and reveal the moral history and destiny of the human race. Man and other living things had been placed 'but a few thousand years upon the earth', and 'the physical monuments of our globe [bore] witness to the same truth'. Mantell saw the extinction of animals as evidence of another of the Creator's laws – that certain races of living beings had been created to suit particular conditions of the earth. When those states became unfavourable, races disappeared and were probably succeeded by new varieties of life. On the subject of disappearing races, Mantell listed the slaughtered fur seals of South Georgia, and the *Apteryx australis*, an 'extraordinary bird' known then only by the single stuffed specimen in the collection of Lord Stanley. Mantell noted that the dodo had been annihilated and 'become a denizen of the fossil kingdom, almost before our eyes', a fate that might easily befall the kiwi. Another high-profile extinction was the Irish elk, whose skeleton was on prominent display in the Hunterian Museum.[6]

By 1839 Gideon Mantell had moved to a house on Clapham Common, London, and was now much more handily located to the several learned societies whose meetings he attended. He had been elected into the Linnean Society in 1813 at the age of 23, and the Geological Society five years later.[7] Walter Mantell sailed for New Zealand in September 1839, leaving his father to his miseries. But this separation proved extremely beneficial, especially to science, for Gideon would soon receive the first of those several major consignments from New Zealand. These would be the subjects of further papers and much discussion at Gideon's scientific meetings, and would also bring him into regular contact, and conflict, with his rival Richard Owen.

Gideon was on the move again, and by 1844 was living at 19 Chester Square, just south of Buckingham Palace Gardens. In mid-year he received a letter from his son, suggesting that New Zealand's 'colossal race of birds' might not be extinct after all. Walter was hopeful of obtaining specimens through a German naturalist residing on the East Coast, and had also heard that they occurred in large quantities on the site of an ancient and deserted Maori pa near Taranaki.[8] If this was the 'interior' of the country he had referred to earlier,[9] it proved extremely fruitful as far as moa bones were concerned. The hoard from Waingongoro arrived at Chester Square in December 1847, and was unpacked and spread out on tables in the dining room. Distinguished visitors invited to inspect the new arrivals included Charles Lyell, William Buckland and Richard Owen.[10] The Hunterian Professor pronounced the osseous assortment 'very marvellous',[11] and his subsequent study also led to his announcement of *Notornis mantelli*. He called on Mantell on at least one further occasion when he and Buckland selected a series of the bones for the British Museum.[12]

Proud owner Gideon Mantell capitalised on his new acquisitions, giving talks at both the London Institution and the Geological Society.[13] When he spoke on the fossil birds of New Zealand to a 'well-attended' meeting of the latter in early 1848, Charles Darwin was there and 'expressed himself much gratified'. But if Mantell was hoping for some rigorous scientific debate, it was marred by William Buckland, who indulged 'more than usual in buffoonery' and the discussion was 'utterly unworthy of the subject'.[14] As well, Mantell may have held the Dean of Westminster responsible for an unsatisfactory monthly meeting of the Archaeological Institute the previous year. Buckland was in the chair, presiding over a very poor attendance and 'still worse affair; mere twaddle on tumuli, barrows etc'.[15] It might be twaddle now, but these ancient landmarks had served the young Gideon Mantell well; the *Dictionary of National Biography* describing him as a 'zealous antiquary, opening many tumuli about Lewes'.[16] Those were the days before systematic archaeology, and in the opinion of another biographer, Mantell 'robbed many prehistoric barrows ... in order to enrich his collection'.[17]

When Richard Owen presented his findings on the bones collected by Walter Mantell at Waingongoro to the Zoological Society on 11 January 1848, Gideon was present. Although it included his announcement of the *Notornis*, Owen's communication was a seriously scientific affair, burdened with such technical terminology as supaoccipital, basi-occipital, eustachian and hypoglossal. For anyone who didn't follow, Owen referred them to his earlier published *Report on the Homologies of the Vertebrate Skeleton*. He described the *Dinornis* as having a 'crocodiloid cranium' and a beak that was used in association with the feet 'in the labour of dislodging the farinaceous roots of the ferns that grow in characteristic abundance in New Zealand'. He concluded by acknowledging Mantell for the privilege of examining this collection, and praised the valuable efforts made by that 'eminent geologist's intelligent and enterprising son, Mr Walter Mantell'. Gideon then offered an opinion, but not on any of the minute detail Owen had just provided. Instead, he suggested that the specimens arriving from New Zealand might not be as recent as had been believed, but might belong to a period of 'high antiquity', along with such other well-known extinctions as the Irish elk and mammoth of England.[18]

In March 1848 Gideon was advised that the Trustees of the British Museum would purchase Walter's collection for £200, payable in June, although it was September when he personally went to Bloomsbury to uplift the money.[19] The following month he took the 10 o'clock express train back to Brighton and spoke on the bones to a 'very respectable but small audience'.[20] But if he had hoped to make a profit from his lecturing it proved an unreliable business, as when only about 50 persons turned up to an afternoon talk he gave to the Marylebone Institution in March 1848.[21] In sharp contrast to this lack of interest on the part of the public, Mantell also had to endure the excessive attentions of Richard Owen. Whilst reading a paper on belemnites – ancient marine animals – to a full meeting of the Royal Society, he weathered a 'most virulent attack' from Owen, who ridiculed his communication. Implying there was a scientific hierarchy, the arrogant Professor stated that Mantell's offering was 'only fit for a few lines in the *Annals of Natural History*'. The two antagonists' colleague William Buckland reportedly 'corroborated' Mantell's views, while the Marquis of Northampton 'passed a warm eulogium' to which the meeting responded.[22] Spencer Joshua Alwyne Compton, the 2nd Marquis of Northampton, was President of the Royal Society from 1835–48. It was presumably in acknowledgement of his personal support that in May 1847, following the discovery of further bones of the *Iguanodon*, Gideon Mantell had announced that he proposed to name a new family of that gigantic herbivorous reptile *Regnosaurus Northamptoni*.[23]

But Mantell would be affronted on further occasions, as when he received a letter from Owen requesting information on this new specimen and his reasons for designating it as he had. He could only console himself in his journal: 'The designing effrontery of the request is too obvious even to me. I must avoid this man in future: but it is very sad thus to be compelled to become as reserved and selfish as the character I despise.'[24] His estimation of Owen dipped further when Dr John Rule called to see his 'Moa's bones', and the two discussed the individual who had perplexed and bothered both of them. Shortly, Mantell received a letter from a collector in Horsham whose fossil specimens from the Weald were about to be the subject of another of Owen's papers, to be published by the Palaeontological Society. 'What next?' Mantell despaired.[25] In December 1848 he went to Horsham to inspect the fossils at issue and, perhaps not surprisingly, the zealous owner would not allow him to make any drawings or notes of specimens. However, Mantell managed to derive some satisfaction from noting 'the imperfect data on which many of Prof. Owen's determinations of the Wealdon reptiles were founded'.[26]

Poor health was another source of constant vexation for Gideon Mantell. In 1841 he had been involved in a carriage accident which caused paralysis of his lower limbs, a tumour on the side of the spine, and severe neuralgic pains in the legs. He struggled on despite the discomforts, at a time when the health of much of London was at risk for other reasons. One particularly cold and damp night in November 1848, whilst returning home along Bird Cage Walk from an Archaeological Institution meeting, he described the stench from the drains by Victoria Road near Buckingham Palace as 'most insufferable'. Later that evening he suffered spasms in the legs, thighs and stomach, as well as extreme coldness; and despite a cocktail of potions that included hot brandy, ether, laudanum, camphor, a hot-air bath and chloroform, the attack did not let up for 48 hours.[27] Meanwhile, cholera had broken out in London. Even William Buckland had 'a touch', in the form of violent diarrhoea, but soon recovered. Later, Mantell went to Oxford to inspect Buckland's Museum, and was disappointed to learn it had aroused little interest. And upon hearing that Buckland's recent class had attracted only six or seven students he despaired that 'there is no hope for natural science in England this century'.[28]

Nevertheless, Mantell persisted in promoting science through his own programme of talks and lectures. In February 1849 his topic 'On a Frog and a Pebble' attracted some 500 persons to the Clapham Athenaeum, with another 100 unable to get in.[29] He also gave an evening lecture at the Whittington Club in the Strand on 'The Colossal Birds of New Zealand', and talked at an 80th birthday celebration, supported with feet, bones and diagrams of the moa. At another soirée, 10-foot high drawings of the moa were attached to curtains of one of the windows in the drawing room, while bones were displayed on a table. Mantell came away from the last engagement 'dreadfully fatigued', but felt 'it was worth the trouble to gratify so many intelligent persons'.[30]

He also kept an eye on the competition, and in February 1849 attended Owen's lecture 'On the Nature of Limbs' at the Royal Institution. Afterwards Mantell had the satisfaction of being able to report that it was far too technical and 'too transcendental' for any audience to understand. Further, it had been advertised to be published before it was delivered – 'What quackery!' When

Mantell's paper 'Additional Observations on the Osteology of the Iguanodon and Hylaeosaurus' was read at the Royal Society in March 1849 there was a full attendance, but Owen was conspicuously absent. The Professor also missed the opportunity to see a table in the library covered with specimens, of which Mantell recorded with satisfaction that 'almost all the fellows inspected them'.[31]

Later that year Mantell learned that the Geological Committee of the Royal Society had decided against awarding him the coveted Royal Medal that he felt was due to him for his work on the *Iguanodon*. He had no doubt this was the result of Owen's influence, confiding to his journal: 'What a pity that a man of so much talent and acquirement should be so dastardly and envious.' He wrote to his ally, William Buckland, who promised to 'do everything in his power to amend the decision'. Justice appeared to prevail, and on 30 November Mantell went to Somerset House to receive his medal. He had the added satisfaction of being able to report that Owen sat opposite him and looked 'the picture of malevolence'. In his own words, he had defeated 'the machinations' of the Professor. A week later he attended a meeting of the Royal Society and received congratulations from all – except Owen who, once again, was not present.[32]

But the state of his health and the professional attacks by Richard Owen were not Gideon Mantell's only concerns. In August 1848 he went to the Covent Garden Theatre, hoping to see the Opera *Les Huguenots*, and was refused admittance – 'with great insolence' – because he was wearing a pair of light-coloured trousers instead of the regulation black. As Mantell noted huffily, this outfit had proved perfectly adequate for the Marquis of Northampton's soirées.[33] Artists could also be problematic, for in January 1849 Mantell engaged the services of George Scharf, who a decade earlier had drawn the first illustration of a moa bone for Richard Owen. Mantell now required drawings for a paper he was preparing on the *Iguanodon*. Scharf was the third artist he had used and, like the others, was proving troublesome.[34] Mantell suffered another personal blow in 1850 when Sir Robert Peel – 'the only eminent public man who payed any respect to art and science, apart from public policy' – died as a result of being thrown from his horse.[35] Two years earlier Mantell had dined at Peel's house, among a party of 16 gentlemen that included fellow scientists William Buckland, Charles Lyell and, of course, Richard Owen.

However, on 6 November 1849 two further boxes from New Zealand brought, among other things, the pair of moa feet which Walter Mantell had collected at Waikouaiti thanks to the observant whaler Thomas Chaseland. For the mounting of these – which probably meant attachment by wire to a slab of timber – Gideon sought the services of the articulator, William Flower.[36] The latter's career had followed Richard Owen's, for he succeeded Owen as Conservator of the Hunterian Museum from 1861 until 1884 and then as Director of the Natural History Museum. But Flower did not always follow Owen philosophically, for he was a friend and follower of Charles Darwin, and when he took over at South Kensington he ensured that the displays reflected the new theory of evolution.[37]

Gideon Mantell now proudly displayed his newly assembled moa feet at scientific meetings around London; and at one of these, at Somerset House, his nemesis arrived 'loaded with bones'. The Hunterian Professor commented in his 'usual deprecating manner' on Mantell's arrangement, claiming it was imperfect. Further, he insisted that the bird had hind toes, and that he had a perfect foot which he described only the *night before* at the Zoological Society! Mantell put this latest attack down to pure envy, and in regard to the supposed hind toe reminded the Society of an earlier 'egregious blunder' made by Owen when he'd mistaken the end of a humerus for the tarso-metatarsal of a wader. But Mantell left the meeting wondering whether it was worth taking 'so much trouble for so little purpose?'[38] Yet undeterred by

Owen's criticisms, he continued to take his assemblage with him, and in June 1850 reported arriving home at 11 pm 'with [his] Moa's feet quite safe'.[39]

Owen's own moa toes had come with a 'magnificent series of remains' which had been collected for him chiefly by the late Colonel Wakefield and sent on by a director of the New Zealand Company, Mr J. R. Gowan. He suspected that a slight depression on the metatarsus – the bone between the ankle and the toes – suggested the existence of a small back toe, and now claimed he had a specimen of the principal bone of that toe, which he proudly displayed.[40] Unfortunately for Gideon Mantell, modern science has come down on the side of the Professor, the issue of the existence of a hind toe having only recently been resolved.[41]

In October 1850 Mantell was able to show off his own latest arrivals from New Zealand – a box of bird-skins which included two species of *Apteryx* and the only known specimen of *Notornis*. He described it as a noble bird, and 'probably the last of the race or nearly so'.[42] A few days later he lent the unmounted specimen to illustrator John Gould to draw for the supplement to his book *Birds of Australia*, and also arranged to sell it to the British Museum for £25. Shortly afterwards that institution also paid £10 for his specimen of *Apteryx oweni*. Richard Owen congratulated Gideon Mantell on his acquisition of this *Notornis* at a meeting of the Royal Society,[43] and pronounced it as belonging to the same genus as the fossil bones already described by him. What was presumed to be an extinct bird was now declared a living species – a story that made the *Times* newspaper.[44] In the meantime Mantell had the *Notornis* skin stuffed and mounted and enclosed in a glass display case to accompany him on his busy meeting and soirée circuit. On one occasion the case was broken and needed reglazing, although the bird itself was 'not much damaged'.[45]

Their collaborations may have benefited science, but there was continuing conflict between Mantell and Owen. In October 1850

Owen sought permission from the Council of the Royal Society to take reprints from nine of Mantell's lithographs of the *Iguanodon*, claiming that the latter's figures were from specimens that had been described by him. Mantell retorted: 'The poor man must be demented.' He pointed out Owen's injustice and misrepresentation in a letter to the Council, and the miscreant Professor made a 'jesuitical apology, and was thoroughly humbled; not a shadow of reasonable excuse could he offer'. It was a satisfying day for Mantell, for that evening he went to a lecture by Michael Faraday on the magnetic property of oxygen, where a lithograph of the *Notornis* was on display.[46] But when he attended a meeting of the Royal Society in May 1851 to discuss the publication of Owen's memoir on the Megatherium, he heard that it was to cost £500 and the proposal was to apply to the Government for a further £250. Mantell was a lone opponent to such a move, 'strongly disapproving' of what he described as 'jobbing'. A week later he attended a further meeting, at which Owen 'demanded' another £100. Mantell left the meeting early, 'heartily sick of the shameful affair'.[47]

In December 1851 Gideon Mantell went to inspect a live kiwi, which had just arrived at the London Zoological Gardens. It was a specimen of *Apteryx mantelli*, recently known as *Apteryx australis* and renamed in honour of Walter's contribution to New Zealand natural history.[48] Gideon found the bird asleep, coiled up in a ball with its beak between its legs, and when disturbed it stood upright and kicked violently, uttering a 'hissing grunt'. [49] The following month Mantell took tea with a Mr Evans, who had sailed on HMS *Acheron* during its survey of unexplored parts of the South Island of New Zealand. On the west were peaks intersected by vertical chasms which extended for miles into the country, and according to natives and whalers, a race of wild men, moa and the *Notornis* inhabited recesses in this region. Over a decade earlier, in his *The Wonders of Geology*, Mantell had hinted at the likely extinction of the kiwi, and now had further grounds for believing this when Evans informed him how he and his party had killed some 30 or 40 of the birds.[50]

The last lecture Gideon Mantell delivered was at the Clapham Athenaeum. Afterwards he took a dose of opium to alleviate his constant pain, but it proved too much for his feeble frame and he died on 10 November 1852. The carriage accident 11 years earlier had caused a debilitating spinal disease, compounded by paralysis of the lower limbs resulting from stooping. Mantell was buried in his home town of Lewes, although part of him remained in London. He had become a Member of the Royal College of Surgeons in 1841, and three years later was enrolled as an honorary fellow.[51] His connection with that institution continued, for his spine was preserved and exhibited at the Hunterian Museum. It demonstrated 'a severe degree of lateral curvature in the lumbar region, with some secondary arthritis'[52] until it was lost during bombing damage to the College during World War Two. In death, as in life, Mantell was at the mercy of Richard Owen, on this occasion in his capacity of Conservator.

Gideon Mantell's death in 1852 denied him his rightful place at one of the most unusual gatherings of scientists ever held. As with the moa, Richard Owen had made dinosaurs his own, and in 1853 was given the job of supervising construction of a series of life-sized models of extinct animals to grace the new Crystal Palace gardens at Sydenham. Designer Benjamin Waterhouse Hawkins and his team created the beasts out of clay – some weighing in excess of 30 tons – which Owen checked for accuracy. Moulds were then made, and from these the dinosaurs were cast. Prominent among them was the *Iguanodon*, the beast that Mantell had first confronted some 30 years earlier. Although ambitious, it was lacking in accuracy, and Mantell would not have been impressed by the size of the forelimbs Owen had now given it. To publicise this undertaking, 21 distinguished guests were invited to a lavish banquet inside the unfinished *Iguanodon* on New Year's Eve 1853. Attached to an enclosing linen awning were plaques recording the names of men who had laboured in the cause of such monsters: Owen, Cuvier, Buckland and Mantell. Beneath them, eleven dignitaries sat down to their eight-course meal at the top table within the beast's belly, while the other ten were accommodated alongside.[53] Mantell's restless spirit no doubt cast a critical eye over the anatomy of 'his' *Iguanodon*, now made concrete.

For a while the late geologist's mortal remains shared shelf space with those of his old colleague and supporter, William Buckland. The latter, his senior by six years, was born in Devonshire in 1784, the son of a rector. Like Mantell, he had become a collector at an early age, stimulated by the discovery of ammonites in rocks near his home. Buckland would continue what had been his father's calling, and combine it with the new pursuit of science. He studied at Oxford and was ordained a priest, and developed his interest in natural phenomena, minerals in particular, with his colleague William Broderip. Using William Smith's geological maps he scoured large areas of south-west England for specimens, accumulating his own geological museum, which he later gave to Oxford University. In 1813 he was appointed to the chair of mineralogy at Oxford University, and five years later to its first readership in geology.[54] It was around this time that the colourful Buckland made an impression on young divinity student, William Williams.

The new science of geology appeared at a time when traditional views on the age of the Earth were coming under threat, and Buckland was able to bring comfort to those who wished to have the Bible supported by scientific investigations. In 1823 he published *Reliquiae, or Observations on the Organic Remains Attesting the Action of a Universal Deluge*, in which he argued that the ancient remains of animals provided a means of judging the inhabitants and character of the Earth before the great flood as recorded in the books of Moses. Shortly afterwards, in a demonstration of the handy accommodation of Church and science – in what was termed Natural Theology – Buckland became both President of the Geological Society and Canon of the Cathedral of Christ Church, Oxford.

In 1836 William Buckland had another high-profile opportunity to advance his interpretation of the Universe. The Bridgewater Treatises were intended to use science to prove 'The Power, Wisdom, and Goodness of God as manifested in the Creation', and Buckland's contribution was *Geology and Mineralogy Considered with Reference to Natural Theology*. He acknowledged the original geologist with a psalm: 'Thou Lord in the Beginning hast laid the Foundation of the Earth.' God was 'their infinite Workman'; evidence of His foresight was everywhere, and no more so than in the very existence and location of coal. Long ago vegetable remains had been buried and converted into this useful mineral, and subsequent changes of level had elevated and made them accessible to fuel-hungry humanity. Buckland noted the advantage of trough-shaped deposits of this carboniferous strata, whose rims were available for mining. He also saw great forethought in the placing of iron ore near the two fuels required for its reduction – coal and limestone. This was such a convenient arrangement that it could have only been done with Earth's present inhabitants in mind. Buckland believed that: 'All the great geological phenomena ... were conducted solely and exclusively with a view to the benefit of man.'55

Coal was now the vital fuel, given Britain's recent conversion to steam power. Buckland estimated there were 15,000 steam-engines at work daily, with one in Cornwall said to have the power of a thousand horses. Steam was to be seen at work 'on the rivers, on the highways, at the bottom of mines, in the mill, in the workshops'. All across the land it 'rows, pumps, excavates, carries, draws, lifts, hammers, spins, weaves and prints'. Such industry was just one example of the economy, order and design provided by the Great Architect. Some of His other achievements were not so easily explained, as when the growing knowledge of the fossil record appeared at variance with the Biblical version of historical events. In attempting to reconcile science and religion Buckland suggested that by placing 'the Beginning' at an indefinite distance before the first of the six days of the Creation, those 'days' might not need to be extended beyond a natural day to allow for 'the necessary preparation of the earth as the abode of man'.[56]

With his great knowledge, enthusiasm, and ability to explain geological phenomena in everyday language, Buckland did much to stimulate the popular interest in the physical and natural sciences. He was a highly original investigator, examining coprolites (fossilised excrement) to determine the ancient meals of reptiles, and studying snails to explain holes bored in limestone. He enclosed toads in artificial cavities to test their tenacity for life, and made hyenas crush ox bones to see if these matched ancient evidence. And when describing fossil footprints he was likely to prance about like the beast concerned, moving more serious onlookers to suggest that science was being reduced to buffoonery.[57]

In 1845, on Sir Robert Peel's recommendation, William Buckland became Dean of Westminster. Three years later he received the Wollaston Medal, the highest honour in geological science, as Gideon Mantell and Richard Owen had before him. When he died in 1856 his son Frank, a surgeon, arranged a post-mortem and noted that while the brain itself was perfectly healthy, there was an advanced state of decay in the base of the skull and the two upper vertebrae of the neck. Today, a small piece of the Dean of Westminster remains on display in a spirit jar in the Hunterian Museum, as an example of tuberculous caries of the cervical spine. According to the label, son Frank was unable to suppress the Bucklandian wit, surmising that his father's terminal condition had not been helped by 'continuous and severe exercise of the brain in thought'.

William's son, Frank (Francis Trevelyan) Buckland was born

in 1826 at Oxford, whilst his father was Canon at Christ Church, and grew up in a home that was a regular menagerie of beasts, with cages of snakes and frogs in the dining room, and where guinea-pigs ran free. As an undergraduate at Oxford, Frank's bedroom became a museum, and when one particular pet became troublesome the master of the college advised him: 'Mr Buckland, either you or your bear must go down.' After gaining a BA he studied medicine in London and was appointed assistant surgeon in the Life Guards in 1854. He took every opportunity to pursue his interests, leading to a series of books, *Curiosities of Natural History*, published from 1857 to 1872. These reflected his eager curiosity and powers of observation, as well as the great breadth of knowledge that caused him to wander off in all directions.[58]

Gideon Mantell had seen the history of creation mirrored in a pebble, but Frank Buckland preferred to contemplate something larger and more animate, such as a horse-pond. Apart from 'mud, dead dogs and cats, and duck-weed' he found such bodies of water contained a world of frogs, shrimps, crabs and newts, which he described with regular recourse to specimens in the Hunterian Museum. His scientific methods were in the spirit of John Hunter, as demonstrated when he determined to find what tadpoles might eat if denied their usual food. By experiment, on subjects obtained from Clapham Common, he found they tended to dine on one another – much as he suggested New Zealanders did after they had eaten 'every specimen of the Dinornis' – 'A huge running bird, otherwise called the Moa.'[59]

The Buckland family is both ancient and extensive, so it may not be surprising that it claims other strong ornithological links with New Zealand. William Buckland (b. 1695) of Crediton, Devon, was the common ancestor of his more famous namesake the Dean of Westminster (1784–1856), and another branch that took the family to the Antipodes.[60] A later William (Thorne) Buckland (1819–76) arrived in New Zealand via Australia in 1841, and three generations later his great-grandson, Geoffrey Buckland Orbell, rediscovered a bird which had not been reported or sighted for half a century — the supposedly extinct flightless takahe or notornis, *Notornis mantelli*. Encouraged by Maori reports of sightings in the largely unexplored area to the west of Lake Te Anau, in the south-west of the South Island, Orbell organised an expedition in April 1948. Descending a valley to an unmapped lake, the party heard unfamiliar bird sounds and saw footprints. They returned to the spot later that year to discover a colony of the elusive birds.

North-east of Hobart in Tasmania is the rural village of Buckland, named in 1846 by Governor Sir John Franklin after his scientific colleague the Dean of Westminster. On the other side of the Tasman Sea, Julius von Haast named the Buckland Peaks in the Paparoa Range, with the Buckland Creek nearby. Some 50 km to the east, across the Inangahua River and the Brunner and Victoria Ranges, is the prominent 1606 m peak of Mount Mantell, named after Walter. And about 500 km further south, to the west of Lake Te Anau, is a living bird that connects those two big names in geology and the discovery of the moa — the *Notornis mantelli*.

Richard Owen poses with the bone fragment (in his right hand) brought to him for identification by John Rule in October 1839. Alongside is the skeleton of a moa discovered by goldminers in Central Otago in the South Island of New Zealand in late 1863 and obtained by the York Museum the following year. This photograph could not have been taken earlier than 1873 when Rule's fragment was relocated and presented to the British Museum. *Canterbury Museum*

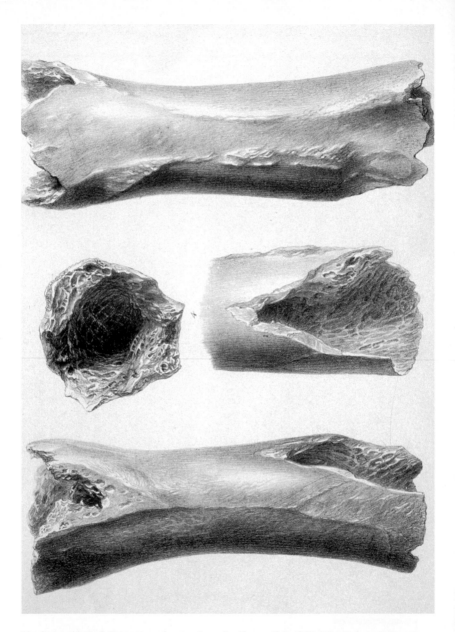

The first published illustration of a moa bone by George Scharf. Dr John Rule took this 15 cm (6 in) fragment to the Hunterian Museum for identification in 1839 and from it Richard Owen deduced that large flightless birds had once lived in New Zealand. These lithographed views accompanied Owen's paper 'Notice of a Fragment of the Femur of a Gigantic Bird of New Zealand', published in the *Transactions of the Zoological Society of London* in 1842. *Reproduced by kind permission of the President and Council of the Royal College of Surgeons of England.*

A section of the interior of a moa femur compared to an equivalent beef bone. Clearly visible on the former is the 'lamello-cellular' or 'cancellate' structure Richard Owen observed – at John Rule's urging – on the original moa bone in 1839. In view of this marked difference in textures, it seems surprising that Owen initially dismissed Rule's specimen as a 'marrow bone, like those brought to table wrapped in a napkin.'

The best-known and most durable of all images of the moa, this interpretation first appeared in Ferdinand von Hochstetter's book: *Neu-Seeland*, in 1863. An English edition, *New Zealand: Its Physical Geography, Geology and Natural History with Special Reference to Results of Government Expeditions to the Provinces of Auckland and Nelson*, appeared four years later. It is probable that this moa was drawn by an artist named Magniani, and based on the moa skeleton discovered in the Aorere Valley, near Collingwood, Nelson, and assembled by Dr Jaeger of the Museum of the Imperial Geological Institution, Vienna.

### INTERIORS AND EXTERIORS. No. 10.

THE MEETING OF THE ZOOLOGICAL SOCIETY, HANOVER SQUARE.

A meeting of the Zoological Society of London in 1885, according to *Punch*. Continuing his commitment to the birds of New Zealand, an 81-year-old Richard Owen (front left) exhibits a hatching apterix (*sic*). He appears startled by Mr Punch's bottled beetle – perhaps a reference to the fact that the young Charles Darwin had been a well-known collector of such insects. In the centre of the picture, William Flower – who succeeded Owen as Hunterian Conservator, Hunterian Professor and Director of the Natural History Museum – deals with a turtle skeleton under the watchful eyes of colleagues. Third from the right is 'Darwin's bulldog', Thomas Huxley, whose presence would add further to Owen's discomfort.

The dodo (above) and ostrich – male and female (right) – from *Rees' Cyclopaedia or Universal Dictionary of Arts, Sciences and Literature,* c.1820. Missionary William Colenso claimed he showed illustrations from this book to Maori throughout the North Island during his inquiries after the moa. *British Museum*

Richard Owen (in gown) explains the fossil skeleton of the mylodon, a large extinct member of the sloth family from South America, to visitors at the Hunterian Museum, c. 1840. The end of the hall is dominated by Chunee, the Indian elephant, with the skeleton of the 'Irish Giant' to its left front, and a giraffe to its right. This view does not include the ostrich skeleton Owen used to identify the first moa bone in 1839. Opposite the mylodon is the fossil shell of a glyptodon, a large extinct armadillo, also from South America. Watercolour by Thomas Hosmer Shepherd. *Reproduced by kind permission of the President and Council of the Royal College of Surgeons of England.*

A corner of the Hunterian Museum, photographed c. 1852. Chunee the elephant overlooks the statue of a negro figure and the bust of surgeon John Hunter. The Museum's growing collection of moa and other large birds is just out of the picture to the left. *Reproduced by kind permission of the President and Council of the Royal College of Surgeons of England.*

The bust of Richard Owen emerges from the rubble following the destruction of much of the Hunterian Museum by German bombing in 1941. *Reproduced by kind permission of the President and Council of the Royal College of Surgeons of England.*

Richard Wolfe, the author, with the 1896 bronze statue of Richard Owen wearing the gown of the Hunterian Professor, on the landing overlooking the Main Hall of The Natural History Museum, South Kensington. Owen was sculpted by Thomas Brock R.A., and holds a moa bone – not the original fragment brought to him by John Rule – in his left hand.

# 10

## IN SEARCH OF NATURAL KNOWLEDGE
### The Scattered Naturalists of New Zealand

*Nothing connected with the natural history of New Zealand has attracted more attention than the extinct moas. Ever since 1840, when Sir R. Owen announced to the scientific world the former existence in New Zealand of a large struthious bird, exceeding the ostrich in size, interest in the subject has not flagged.*

CAPTAIN F. W. HUTTON, 1892[1]

New Zealand's first laboratories were improvised cabins on European sailing ships. Stuffy, confined, and at the mercy of the motion of the ship, these were hardly ideal conditions for the pursuit of science. Such a working environment was recorded by Joseph Banks on the *Endeavour*: 'Dr Solander setts at the Cabbin table describing, myself at my bureau journalising. Between us hangs a large bunch of seaweed, upon the table lays the wood and barnacles . . .'[2] This desk-work naturally followed field-work, which – given the nature of the New Zealand terrain – presented early scientists with other challenges. Such on-board laboratories were not the places for definitive descriptions and dissections, but rather for the preserving and packing of specimens for the safe trip back to Europe. The recently discovered lands in the Pacific provided much raw material for European expertise, but, once organised settlement was under way, New Zealand began to establish a scientific community of its own. In time, the extinct bird that had been the exclusive domain of England's Richard Owen could then be dealt with on its own turf.

Modern science had its beginnings in the 16th century when Europe's horizons were expanded by the new age of exploration. A more systematic understanding of the natural world, from the minuscule to the astronomic, was made possible by such developments as the compound microscope and telescope. Learned bodies began promoting the study of scientific matters, and organising expeditions to bring back specimens. Britain's first such body was the Royal Society of London for Improving Natural Knowledge, founded in 1660; and in 1768, as the much abbreviated Royal Society, it began planning an expedition that would lead to New Zealand.

Eighteenth-century reasoning suggested a great unknown southern land – *Terra Australis Incognita* – existed to maintain the Earth's equilibrium. The British Admiralty was keen to investigate this alleged land-mass, but did not want any such expedition to attract the attention of rival European powers. The Royal Society provided a convenient cover for this venture, for it was sending Lieutenant James Cook to the South Seas to observe the transit of the planet Venus across the face of the Sun. Such celestial events occurred infrequently, in pairs eight years apart and with an interval of over 100 years between pairs, and enabled astronomers to take measurements with which they might calculate the distance of the Earth from the Sun. Observations taken in 1761 had been unsuccessful, and so the next opportunity would come in 1769.

The barque *Endeavour* sailed from Plymouth in August 1768. The following June 3 dawned a perfect day for transit-spotting in Tahiti, with not a cloud to be seen. The observation was made in broad sunlight, with telescopes projecting images of the sun on to a screen. But whereas Cook and his team had expected Venus to appear as a sharp and perfectly circular disc, it produced a blurred edge which created havoc with their calculations. Unfortunately, the worldwide observations that year proved to be of no more scientific value than those of 1761.[3] (The next transit was in 1874, and posted to New Zealand to observe it was a member of the Royal Engineers – Charles Darwin's son Leo.)[4]

His observations may not have provided a key to the size of the universe, but Cook had other matters to deal with. He now followed the Admiralty's secret instructions, requiring him to search for this *Terra Australis Incognita*. If found, he was to explore its coast, along with 'the nature of the soil and the products thereof, the beasts and fowls that inhabit and frequent it'. The *Endeavour* sailed due south, crossing the Tropic of Capricorn and continuing to latitude 40°. As instructed, Cook then turned west, and by late September, encountering mounting evidence that land was nearing, offered a gallon of rum and naming rights to the first person to sight it. At 1.30 in the afternoon of 6 October, Nicholas Young gave a shout from the mast-head and duly claimed the prize.

The *Endeavour* entered New Zealand waters and remained there for the next six months, enabling the first systematic study of the country's natural history, by Joseph Banks and Daniel Solander. Nearly a century later, fellow botanist William Colenso described this scientific event. On the evening of Sunday 8 October 1769, the above pair 'had the pleasure and privilege of beholding and gathering the first floral specimens of (what they then believed to be) the vegetation of the great *terra australis incognita*'. Colenso effused: 'That was truly a Botanical æra; when the queen of natural science (through the efforts of the immortal Linnaeus and his zealous disciples, aided by their royal patrons and promoters), vigorously flourished, and bore those pleasing and useful fruits which have come down with such good results to our own times.' Colenso recalled also how he had been overcome by a 'deep reverential feeling' when he first visited Tolaga Bay and followed in the footsteps of those pioneers.[5]

Those men on the *Endeavour* established a scientific habit that would be maintained on Cook's two subsequent voyages to New Zealand. During the second, on the *Resolution* – by which time Cook had become a captain – at Dusky Bay at the south-western corner of the South Island, Johann and Georg Forster were able to collect many new species of birds which had the misfortune to come within range of these shooters for science.

Following Cook's visits, men of science flocked to New Zealand. Most notable was Charles Darwin, although his brief stay in this country in 1835 had no significant influence on his later development. Six years later the Bay of Islands was a temporary home for the most highly honoured botanist in history, Joseph Hooker, who made significant contributions to New Zealand botany and proved a handy contact for William Colenso. And while many of these early visitors to New Zealand had received some degree of scientific training, they were more likely to come as ship's surgeons than scientists. But irrespective of their official duties, they kept an eye out for specimens likely to be of interest back in Europe.

As the secrecy surrounding Cook's first voyage indicated, it was not just Britain that had an interest in the Pacific. French commander Jules Sébastien Cesar Dumont d'Urville also made three expeditions to New Zealand, spending 67 days in its waters in 1827. At Tolaga Bay, not far from where Joel Polack would shortly encounter the ossified remains of a moa, d'Urville's men found out about a related bird when they saw a Maori cloak trimmed with kiwi feathers.[6] Their reports further increased European awareness of New Zealand bird-life in the Northern Hemisphere.[7]

When the organised European settlement of New Zealand

began, scholarly pursuits were considered essential for a cultivated society. Future residents of Nelson could barely wait, for in
April 1841 they founded a Literary and Scientific Institution
whilst still at sea on the way to their new settlement.[8] It was the
country's first learned society, and by 1844 its membership of 60
included most of the local prominent men. For social balance
there was a less exclusive Mechanics' Institute, which offered a
library along with lectures and elementary adult education.

The first steps to put local science on a professional footing
came with the inaugural meeting of the New Zealand Society, in
Wellington on 2 July 1851. It was no doubt inspired by Sir John
Franklin's successful Tasmanian Society – which had several
New Zealand members – and its interests encompassed the
'physical character' and natural history of the country, along with
the culture and traditions of the Maori. Its activities extended to
a museum, which contained fragments of moa egg-shells and a
collection of rocks and minerals sent to the Society's first honorary
secretary, Walter Mantell, by his father Gideon in England to
encourage systematic geology in the colony.[9] The fact that the
younger Mantell read a paper on the 'various conditions under
which [he] had found remains of *Dinornis* and its contemporaries'
at the Society's first meeting,[10] suggested the moa would become
a major subject for the country's fledgling scientific community.
But Mantell left Wellington for the South Island soon after the
Society was founded, and it lapsed after only its second annual
meeting, in October 1853, failing to respond to several attempts
to revive it.[11]

Another important early development was the establishment
in 1852 of the Auckland Museum, the country's oldest such
institution, which began with a primary aim of collecting
specimens of economic importance to the colony. New Zealand's
scientific circles were then occupied largely by enthusiastic
amateurs, otherwise employed as doctors, farmers, lawyers,
administrators, teachers or clergymen. They operated as part of

an international web dedicated to the advancement of scientific knowledge and the glorification of the British Empire. Sitting at the centre of this network was the Hunterian Professor Richard Owen, and thanks to him the moa would become this country's major contribution to the world of natural history. Owen and his colleagues needed specimens, and the scattered naturalists of New Zealand were happy to oblige their masters back Home.[12]

One of Owen's many regular suppliers was George Edward Grey, who had trained at the Royal Military College, Sandhurst. In 1837 he was granted leave from the army to lead an expedition to Western Australia with the aim of promoting settlement. He sailed on the *Beagle*, bunking down in the cabin Charles Darwin had occupied six years earlier. The fauna of Australia posed a puzzle for scientists of the day, with Grey dispatching specimens back to the Royal College of Surgeons. Owen thanked him for two crates from the Swan River, which shortly stimulated a monograph on the hooded lizard. Owen also took the opportunity to advise of 'what has been doing in Science at home'. Mainly, it was 'Darwin's Zoology', the material collected in South America during the *Beagle* voyage, including six new and gigantic beasts which had 'left their bony remains scattered about the plains of Buenos Ayres or Patagonia'. Owen then turned to Grey's present circumstances, claiming: 'There is no field of Discovery which interests honest John Bull more than his far away possessions in the Australian Wilderness.' To satisfy this curiosity, more skulls and bones were needed, irrespective of condition. Surely no comparative anatomist has expressed the need more colourfully than Owen then put it to Grey: 'Starve the Dingoes; don't let them crack any bones, save the Mutton: box up in saw dust all the odd bones of Marsupiala.' The thorough Owen was especially interested in the 'long-eared soft-furred Rabbit-rat of Swan river', a small ant-eating Bandicoot 'that may be occasionally smoked out of [high?] trees', and was also keen to get hold of the 'impregnated uteri' of any marsupials.[13]

When he was back in England in 1840, the 28-year-old George Grey wrote an influential report on the government of indigenous peoples for Lord John Russell, Secretary of State for the Colonies. This led to his return to the Antipodes, as Governor of South Australia, which in turn was a good apprenticeship for his next appointment, as Governor of New Zealand from 1845. He took up that office at a time of increasing tension between new settlers and Maori, the most serious threat to European authority being posed by Hone Heke and his ally Kawiti in the far north of the North Island.

In November 1845 Richard Owen wished Grey well in those troublesome times, and had faith he could hold out against the insurrection: '[I] entertain so confident a hope of your being able to maintain Auckland against General Heke as to direct this letter to the capital of your present island.'[14] Owen's reference to the 'present island' was less a suggestion that Grey might have had to retreat to New Zealand's South Island than a reflection on his own hazy grasp of that country's geography. Even so, Owen trusted these annoying affairs of state wouldn't get in the way of science. He sent copies of his recently published scientific papers in the hope they might stimulate the collection of more material on the *Apteryx* and *Dinornis*, in particular the head and beak of the latter.[15]

In 1848 Owen wrote that both he and Mrs Owen had 'eagerly perused every article' relating to Grey's 'progressive mastery of the difficulties' he faced in New Zealand, but again stressed the importance of his continuing contribution to science: 'You have given me abundant proof that the more immediate and arduous cares of Government will not divert you from the due encouragement of the collection of the materials for the natural history of your Colony.' In the same letter Owen referred to a collection of remains of extinct great birds of New Zealand which had just arrived from the 'laborious & enterprising son of my eminent & esteemed friend Dr Mantell', a most generous tribute that would surely have come as a surprise to Gideon Mantell.[16]

Grey used his own connections and contacts to seek out the requisite moa remains, and another box reached the Royal College of Surgeons in June 1850. This contained bones from a cave at the base of Mount Tongariro in the central North Island, and a delighted Owen described them as the 'first cavern-specimens which have been transmitted'. He assured the sender that 'since the reception of the first box of specimens sent by Archdeacon Williams in 1843, no collection has been contemplated by me with more delight than the present'. As well as moa, it included 'remains of a large short-winged water-rail (*Notornis*)', which led Owen to suggest it 'may, perhaps, still exist in some parts of New Zealand'.[17] This 'valuable consignment' of cave remains had its first public outing at a meeting of the Zoological Society in November 1850.[18]

In the meantime there had been a calamitous event, the scientific implications of which will never be known. On 23 June 1848, Government House in Auckland, the home of Sir George Grey, was destroyed by fire. This wooden structure had been prefabricated in England and shipped out, and was a counterpart to one built 20 years earlier for Napoleon's use at St Helena. The fire destroyed a large amount of Grey's household and personal items, including natural history specimens. The extent of the latter loss was made apparent in a newspaper item stimulated by *The Governor*, a New Zealand television drama based on Grey's colourful life.[19] This referred to a letter from Grey to the writer's great-grandfather, Richard Owen, describing material which had been destined for the Professor. Lost to the flames was 'a magnificent collection of moa bones, including a complete skeleton of the largest moa which had ever been found' and three complete and extremely rare moa heads. There were also moa bones which had been gnawed and 'crunched' by some large animal, and still bore teeth-marks. As if this wasn't enough, Grey mentioned he'd also had specimens of a new bird, 'allied to the kiwi' but much larger, and said to be extinct. But the bones that might have caused the most interest in zoological circles were those Grey saved mention of for last. He was 'almost afraid to say it', but he claimed to have had bones which 'we all regarded as the rudimentary wings of the moa'. In spite of this setback, Grey ended on a cheery note, suggesting that during the summer he would 'collect again as much as I can'.

Owen was most diplomatic in his response, expressing his and his wife's distress at this 'grievous calamity by conflagration'. But the loss of these 'noble' bones was 'insignificant in comparison with that which you have sustained in the destruction of property including so large a proportion which Europe only can supply to the resident in a new Colony' – no doubt a reference to the destruction of much of Grey's china, linen, wine and library in the blaze.[20]

George Grey's network of correspondents included Charles Darwin, who wrote in late 1847 explaining that – quite unlike Owen – he had 'no personal interest' in any points of natural history in New Zealand. Nevertheless, he did mention the limestone caverns in the Bay of Islands ('& I daresay elsewhere') which he had been prevented from entering because they were used as Maori places of burial. He suggested that 'digging in the mud under the usual stalagmite crust would probably reveal bones of the contemporaries of the Dinornis'. Darwin also inquired whether 'erratic boulders' existed in New Zealand, and closed by apologising to Grey for 'providing so much geological detail' and suggesting – somewhat unfortunately, as it turned out – that he could 'burn this letter'.[21]

Auckland's new Government House was completed in 1856, but too late for George Grey. He had left the country at the end of 1853 to become Governor of the Cape Colony and High Commissioner for South Africa, and until then lived in temporary quarters. Whilst there he received a visit from one Thomas Shearman Ralph, a Member of the Royal College of Surgeons, who Owen recommended for scientific work in the colony.[22] It seems that Grey was able to oblige, for Ralph shortly became honorary secretary for the New Zealand Society, and it was in this capacity that in April 1852 he could advise Richard Owen that he'd been elected a member of that same body.[23] Ralph's contribution to local science included a paper 'On the Katepo [sic], a supposed poisonous spider of New Zealand', published by the Linnean Society in London.[24] He later emigrated to Australia, and George Grey was also on the move. In London in 1854 he presented 142 items from his collection to the British Museum, among them a Maori greenstone adze described as having been 'injured' in the fire at Government House in Auckland.[25]

Botany was the first area of New Zealand natural history to be studied on site by scientists from Europe, to be followed by the younger discipline of geology. The first serious studies of the foundations of New Zealand were carried out in the interests of the national economy, and quickly led to the moa. Once again, George Grey had a hand in this development. Whilst he was in South Africa, the Austrian frigate *Novara* – those being the days when the Austro-Hungarian Empire extended to the Mediterranean and possessed a navy – called in at Cape Town during the course of an extensive scientific expedition. Grey encouraged the commander to visit Auckland, to study the volcanic regions of the North Island. He followed Grey's advice, and after calling at Ceylon (Sri Lanka), India and Sydney, and sustaining damage from a hurricane in the China Sea, the *Novara* reached Auckland on 22 December 1858.[26]

The population of the 18-year-old settlement of Auckland was less than 8000, and centred on a small bay at the bottom of a mostly muddy thoroughfare known as Queen Street. To its immediate east lay Point Britomart, the promontory that was to be demolished from the late 1870s to provide for reclamation of part of the Waitemata Harbour. In 1858 the *Novara* was the largest warship yet seen in Auckland, and on board was a German-born geologist, Ferdinand von Hochstetter. Whilst in London a few weeks prior to the departure of the ship he had marvelled at the moa display in the British Museum.[27] Aware that the expedition planned to visit the South Pacific he was hopeful that he might be able to obtain such specimens himself. The day before he stepped ashore at Auckland, the *Evening Star* had berthed with a second German geologist, 36-year-old Julius Haast, who had come to New Zealand to study the prospects for German emigration, and who would also become involved in the quest for the same large bird.

Hochstetter was asked by the Provincial Government of Auckland to report on a local coalfield, and following this was persuaded to remain in New Zealand and carry out further geological survey work. While the *Novara* sailed for home without him, he and Haast, along with other assistants including a retinue of twelve Maori porters, spent some nine months travelling throughout the North Island. But apart from obtaining two single bones, one being a pelvis given to him by a Maori chief who had hidden it among the 'dust and rubbish of his raupo hut', Hochstetter was unsuccessful in his hunt for moa specimens. He 'scoured every district' known to have had such bones and admitted also that he 'ransacked all the so-called Moa-caves', but 'in vain'. Others had beaten him to it, and local Maori were now well aware of the monetary value of such items and had gathered up all they could find to sell to 'European amateurs at enormous prices'.[28]

Haast and Hochstetter were the first of four distinguished geologists to come to New Zealand within the next decade, the others being James Hector and Frederick Hutton. In addition to the coincidence that all four surnames began with 'H', their individual career paths would also lead to the moa. The two Germans quickly established a lifelong friendship, which would later be continued by their widows. Hochstetter also named his first daughter Julie, after Haast, who acted as her godfather.[29]

In August 1859 the pair crossed to the north of the South Island, where they prospected for copper and coal, and Hochstetter was finally able to obtain some evidence of the bird that had so far eluded him. Haast spent three days and three nights excavating and washing moa bones found in a cave in the Aorere Valley, and they were then brought back in triumph to Collingwood, on bullocks decorated with flowers. The whole town turned out to greet the arrival of the beasts bearing the bones. Among them was a near perfect skeleton, which was passed on to the Nelson Museum, and whose trustees in turn presented it to the Museum of the Imperial Geological Institution at Vienna. At last, Hochstetter would be able to return to Europe with evidence of New Zealand's 'perished bird world'.[30]

Some two months after that moa discovery, Hochstetter sailed for Europe. He never returned to New Zealand but he and Haast remained regular correspondents and assisted each other's endeavours. Haast undertook further explorations for the Nelson Provincial Council, and later became Provincial Geologist in Canterbury. In 1862 his now sizeable personal collection of specimens was supplemented by the arrival from Vienna of a plaster-cast of the restored moa skeleton that had been found in the Aorere Valley. The bird's growing popularity was reflected by the fact that another 29 casts were made available for sale by the museum in Vienna to other institutions around the world, at 100 florins each.[31] That same year Haast initiated a meeting to found the Philosophical Institute of Canterbury, and suggested some objectives for the body. Second on his list, after the establishment of an acclimatization society for the introduction of 'useful plants and animals', was the 'erection of a Museum of economic geology and of natural history generally'. Here, 'Collections should rouse a spirit of observation in all' and 'every rock, plant and animal, that might otherwise be lost, might find a place'.[32]

Before long, the draining of a swamp at Glenmark, North Canterbury, proved a stimulus for establishing the very museum

Haast had in mind. Alerted by one of the owners of the property, Mr G. H. Moore, Haast hurried to the site and saw a spectacle that – as one writer put it – 'must have thrilled his scientific soul'. Moore provided workmen to assist with the excavation, and Haast returned to Christchurch with a horse-drawn wagon laden with moa bones. He estimated the swamp contained the well-preserved remains of at least one thousand such birds, and in spots they were so closely packed together they could only be extricated with difficulty. Glenmark was a gigantic graveyard whose bones lay in 'heterogenous confusion', and over 60 years later was described as 'the most extraordinary [find] of its kind which so far has been made, or is now ever likely to be made'.[33]

In his capacity as Provincial Geologist, Haast had rooms in the local Government Buildings in Christchurch, where bones from Glenmark were carted for sorting and assembling. From these specimens, which included also the remains of a previously unknown giant extinct eagle, seven moa skeletons were assembled and put on display. The success of this venture was an incentive for the local government to establish a separate museum, and Julius Haast was appointed its founding director. The Canterbury Museum opened to the public on 1 October 1870, starting out as a single lofty room with an upper gallery supported on wooden columns. Predictably, its central display consisted of a cluster of some eight or ten moa skeletons, the largest, *Dinornis maximus*, standing an impressive 11 feet 7 inches. For comparison, they were shown alongside ostrich, emu and human skeletons, and also that of Jack, the local Acclimatization Society's late kangaroo.[34] The skeletons raised from the swamp at Glenmark had another ongoing benefit for the new museum, for Haast was able to exchange them with institutions overseas and further extend his collection.

By now, Ferdinand von Hochstetter had been back in Europe for a decade, and was working on material collected during the *Novara* expedition. This resulted in 21 scientific papers on

New Zealand and the publication in 1863 of his *Neu-Seeland*, the first substantial book on the country in the German language.[35] Hochstetter's old friend Haast was instrumental in getting the New Zealand Government to subsidise a translation, and an English edition appeared in 1867. In his preface Hochstetter acknowledged the assistance of his 'friend and former fellow traveller Dr Julius Haast', and hoped that his fellow countrymen would be 'gratified' by his presentation of the 'wonders and peculiarities' of that southern land. He also quoted the noted German geographer Carl Ritter, who believed New Zealand was destined to become a 'mother of civilized nations'.[36]

Among his illustrations of the wonders and peculiarities of New Zealand, Hochstetter included its large extinct bird. Although there had been images of the bird in scientific publications, this was probably the first to enjoy popular circulation and establish its public image. In the first instance this image was due to Hochstetter's colleague, Dr G. Jaeger, at Vienna's Imperial Geological Institution, who had laboriously restored and made plaster-casts of the *Palapteryx ingens* skeleton retrieved from Aorere Valley. It was short of a few original bones and its pelvis was modelled on that of a different species Hochstetter had collected in the North Island. The completed skeleton stood $6\frac{1}{2}$ feet (1.98 m) to the top of its skull and was considered a 'masterpiece', a credit to Dr Jaeger. Hochstetter also acknowledged an artist, Mr Magniani, who drew two views of the assemblage for his book *Neu-Seeland*. Magniani was most likely also responsible for the now famous rendition of the complete bird – along with three kiwi – in its natural habitat, and its final published form was probably engraved by Eduard Ade.[37] The image proved extremely durable, with one version even serving as a trademark for a brand of New Zealand egg preserver until at least the late 1940s. Unfortunately, that product came too late to assist the animal itself, for only 18 complete eggs, or broken eggs able to be usefully pieced together, are known to science.[38]

The popularity of 'Hochstetter's' interpretation of the moa was due largely to the unashamed height it gave to the bird. Like the museum reconstructions that it no doubt inspired, it made the most of the moa's potential. With neck fully erect, it towered over scrub and fern, against a rugged mountainous backdrop, while three minuscule kiwi foraged in its shadow. But for all the exaggeration, Hochstetter himself was 'not inclined to believe' some of the stories then circulating about recent last moa sightings, and assertions that American sailors and sealers had seen 14-foot, 16-foot and even 20-foot birds 'stalking to and fro' at Cloudy Bay and other inhospitable parts of the South Island.[39]

Recognising that the moa was the 'principal game of the natives', Hochstetter wondered where they obtained their food after its extermination and, like others before him, identified cannibalism. But in his view it was hunger, 'not barbarity, not savage cruelty, not monstrous heathenism' that had driven 'the uncivilized man of the South seas . . . to drink his fellow's blood and eat his flesh'. As Hochstetter pointed out, Europeans had also been known to do it, when shipwrecked. Cannibalism was in fact 'one of the manifold forms of the struggle of life', and it was the arrival of swine and potatoes with Europeans which put an end to the practice in New Zealand.[40]

From Europe, Hochstetter could report on the arrival of two of the 'wingless' birds featured in his 1863 woodcut. A live 'hen-kiwi' had arrived at the London Zoological Gardens in 1852, and required daily rations of $1/2$ lb. of mutton, plus worms.[41] Also, there was a skeleton of its extinct relative, *Dinornis elephantopus*, at the British Museum. It had been among the 'most copious harvest' of bones collected by Walter Mantell at Awamoa in North Otago, a locality which he also named, in 1852. That massive skeleton had been placed, 'very appropriately', by the side of the gigantic elephant, *Mastodon ohioticus*.[42]

While the moa's profile was being raised in more ways than

one, New Zealand science was undergoing a similar trans-
formation. About to play an important part in this development
was James Hector, who had graduated in medicine from the
University of Edinburgh in 1856. The following year he went as
surgeon and geologist on an expedition to western Canada where
he combined scientific surveying with high adventure, dealing
with warring native tribes and discovering the Kicking Horse
Pass.[43] In 1861 he was appointed director of the Geological Survey
of Otago, New Zealand. His name came to the attention of
authorities in Wellington, who now recognised the value of
systematic scientific study. The need to place this on a firmer
footing led to the establishment of the New Zealand Geological
Survey in 1865, and Hector was appointed its first director. As
such he became the nation's top scientist – and its busiest – for
his raft of responsibilities also included the Colonial Museum,
Meteorological Department, Colonial Observatory, Wellington
Botanical Garden and Patent Office Library. Yet another of his
hats was worn as editor of *Transactions*, the publication of the
recently founded New Zealand Institute. This was the result of
an Act of Parliament aimed at establishing 'an Institute for the
advancement of Science and Art in New Zealand'. At first it
incorporated the Auckland Institute, Wellington Philosophical
Society, Philosophical Institute of Canterbury, and the Westland
Naturalists' and Acclimatization Society, and later the Otago
Institute and others.[44] In the first issue of the new national body's
*Transactions* in 1868, Hector noted the particular importance of
papers relating to the natural history and resources of the colony,
and offered a list of suitable topics for potential contributors. These
ranged from the mythology of the moa to the much grimmer
reality of the sanitary conditions of the colony's cities.[45]

To mark the centenary of Cook's rediscovery of New Zealand
in 1869, Julius Haast made a novel suggestion to the Philosophical Society of Canterbury. Because the largest of the
country's three main islands laboured under the confusion of two

names – Middle Island and South Island – he offered 'Cookland', in recognition of the 'illustrious navigator'. He had seriously considered honouring Dutch navigator Abel Tasman, who came in 1642, but conceded that 'Tasman Land' would be confused with the southernmost state of Australia – which had recently undergone a name change of its own, from Van Diemen's Land.[46]

Soon, Haast had opportunities to place the moa on the world stage. The first occurred when Hochstetter was appointed a member of the Imperial Commission for the International Industrial Exhibition to be held at Vienna in 1873, and wrote to Haast hoping that New Zealand would be well represented. Sensibly, Hochstetter was invited to assist in the arrangement of the New Zealand exhibit, and took a personal interest in the assembly and restoration of its four moa skeletons, three being supplied by Haast and the fourth from the collection at Vienna. They were supplemented by four stuffed kiwi, and all stood on a tall platform covered with red linen. Thanks to Hochstetter's influence the 'Extinct Moas and still living Kiwis' stood in the centre of their Court where they could not be missed, and even attracted the personal attention of Emperor Franz Josef. He was, reportedly, astonished at their size, but others were merely confused, for one member of the jury commented: 'What a fine giraffe that is.'[47]

Eleven years later Haast had another opportunity to take the moa to the world when he was appointed New Zealand Commissioner to the Colonial and Indian Exhibition in London in 1886. His concept for the New Zealand display was one that combined the picturesque with the utilitarian, and included three moa skeletons to represent extinct fauna. Although there were criticisms of the final display – including complaints there was too much natural history – the *Times* concluded that none of the Courts was more 'systematically arranged or more attractive and instructive than that of New Zealand'. The headmaster of the Brighton Grammar School produced a popular pamphlet on the

Exhibition, and described New Zealand as 'another daughter nation, whose aim, judging by her Courts, is resolute, intelligent progress'. Few failed to be impressed by the central moa group, although another confused visitor was overheard remarking that the 'giraffe' was missing its hind legs.[48]

Having successfully raised the moa from the swamp, science now wondered when the bird had became extinct. Haast believed it had been hunted by a race that existed in New Zealand prior to the Maori and were contemporaries of the palaeolithic 'cave men' of Europe. To clarify this issue, in 1872 he employed two work-men to excavate Moabone Point Cave, near Sumner, Christchurch. They turned up a huge collection of bones, artefacts and other items, all of which followed the usual route to the Canterbury Museum. Haast was now engaged elsewhere and unable to publish the results of these excavations, so was infuriated when one of his workmen – Alexander McKay – wrote a paper on the subject which was read at the Wellington Philosophical Society in 1874. Galvanised into action, Haast completed his own paper and succeeded in getting it published before McKay's version went to print. He unkindly described McKay as a simple labourer – in fact, as a 'mere mullock-turning machine examined and cleaned at stated intervals' – even though the latter had forced Haast to change his thinking and accept that moa bones were contemporaneous with polished stone tools. McKay had acted because of what he perceived as Haast's tardiness, perhaps viewing him as sitting in a commanding position and thwarting the efforts of others – much as some regarded Richard Owen's influence in Britain.

By the late 1880s the discovery of distinctly Maori artefacts alongside moa bones in archaeological sites suggested that Haast's belief in the great antiquity of the bird was wrong. It would soon be generally agreed that it was Maori, direct ancestors of the living people, who had seen the last of the moa, between three and five centuries earlier.[49]

In 1891 the first half-century of moa research was summed up in a paper given at a meeting of the Philosophical Institute of Canterbury. It was read by Frederick Wollaston Hutton, who had came to this country in 1866. He had been born in Lincolnshire, into a family with strong scientific connections – the Wollaston Medal being the Geological Society's most prestigious honour.[50] In New Zealand he enjoyed a succession of scientific appointments in Otago and Canterbury, and now began his review of the extinct moa by pointing out that no aspect of New Zealand natural history had attracted more attention. In fact, ever since Richard Owen had drawn attention to it, 'interest in the subject has not flagged'. But knowledge of its structure and history was still, like its bones, 'very fragmentary', and were matters of confusion which Captain Hutton – the title a result of his service in the Crimean War and Indian Mutiny – now hoped to clear up. To this end he had examined most of the relevant collections in New Zealand and also studied the publications, in particular those of Richard Owen, James Hector and Julius Haast. Hutton described all moa as having been 'remarkably robust in build, with strong legs, and rather flat heads with small eyes but well-developed olfactory organs'. The whole bird was covered with soft fluffy feathers, and Hutton recommended Hochstetter's illustration as 'an admirable restoration'. However, some South Island species must have been 'absurd-looking creatures', nearly as broad as high, and waddling on short stumpy legs, and, in general, they were all 'stupid and sluggish'.[51]

Hutton also disputed Haast's belief in the great antiquity of the moa hunters, concluding that in the North Island the moa was exterminated by the Maori not very long after their arrival in New Zealand – not less than 400–500 years ago, and was last seen in the South Island some 300–400 years ago. He understood the birds were descended from ancestors of the Northern Hemisphere, and had spread southwards through New Guinea to Australia and New Zealand. Finding conditions so favourable here, the birds had multiplied and spread, resulting in a much greater number of different species of struthious birds than in any other part of the world. The proliferation of species was put down to changes in the sea level, causing successive periods of isolation to the early New Zealand land-mass, as a result of which the original bird-stock diversified, remingled and multiplied.

Hutton now acknowledged a new scientific concept when he claimed that as far as the moa was concerned, 'natural selection did not come into play'. The living was easy, without carnivorous mammals or other enemies to keep them in check, and no shortage of vegetable food. Under such favourable conditions, the birds simply got 'larger and fatter, more sluggish and more stupid'.[52] Hutton was well aware of Darwin's theory that suggested the 'survival of the fittest', a struggle for existence in which offspring inherit certain traits that better equip them to deal with the ordeals of nature. When carried on by successive generations, such imperceptible adaptations to changing conditions could produce what may be termed a new species. But if Hutton believed the moa had no truck with this new theory, he was not alone.

The idea of progressive development from one species to another had been labelled a 'heresy' by such notable scientific gentlemen as Sir John Herschel and Sir Charles Lyell.[53] All species were then regarded as special creations, and it was therefore extremely presumptuous for science to imagine that it had an explanation for what was surely one of Nature's greatest secrets, the 'very mystery of mysteries'. But by the end of the 19th century, evolution had become accepted as a demonstrated principle, and its chief proponents were regarded less as infidels as providers of an alternative explanation for the wonders of nature and the workings of God. The issue then became whether the mechanisms suggested by Darwin were sufficient to explain this evolutionary process, or whether there were other as yet unknown forces involved.

In 1861, five years prior to coming to New Zealand, Hutton had written a favourable review of Darwin's *The Origin of Species*, for which he was cordially thanked by the author.[54] He remained a staunch advocate of evolutionary theory, believing it explained the first stage of a process that continued beyond the grave. However, his support for evolution on Earth did manage to upset elements of the Presbyterian Church while he was Professor at Otago University.[55] The Anglican Church also had difficulties with such ideas, Bishop Williams dismissing them as the 'machinations of Satan', while Bishop Hadfield warned his clergy in 1876 of 'shallow physical theories which are supposed to be repugnant to what we accept as Divine truth'.[56]

Meanwhile, the learned societies of New Zealand provided opportunities for clergymen and laypersons alike to air their feelings on such contentious matters. Some presentations managed to straddle the divide, and were a long way from the identification of bones from swamps. But the most remarkable ideas relating to evolution were surely those put forward by Alexander William Bickerton, a university professor and eccentric. As a young man he had attended lectures in London by John Tyndall and Thomas Huxley, and in 1874 was appointed Professor of Chemistry at the newly established Canterbury College in Christchurch, New Zealand. Twenty years later the controversial Bickerton was the subject of an inquiry into the management of his department, and received support from a student who became the greatest physicist of the age, Ernest Rutherford.[57] Bickerton also served as president of the local Philosophical Institute, and developed a theory that the principle of evolution went far beyond this Earth and was a universal law with cosmic implications. In his opinion: 'When the perfection of the cosmos is realized, joy will be seen to be the true lot of man . . .'[58] Joy may not have been the lot of certain of the clergy that year, for it was noted that some of their congregations were being lured away by Eastern philosophies. But an accommodation between God and Darwin may have already been under way, for when the Rev. J. Bates spoke to the Auckland Institute on 'Comparative Religion' he mentioned that clergy were 'attempting to bring religion into line with modern thought'.[59]

The moa had not, of course, escaped natural selection. One of the main forces behind that process was the continual and gradual changes to conditions, such as climate and sea-level. The moa was certainly subject to these – even if, as Hutton suggested, it was too stupid to notice – for it was the changing shape of the landscape that set the pattern for the extensive moa family. Another important factor in natural selection is when population growth outstrips food resources, but the moa was probably spared this challenge, on account of its modest breeding efforts – single-egg clutches may have been common among smaller moa, with two eggs laid by the larger species – and plentiful food stocks. Moa could also thank their ancestors for inheriting sturdy ground-scratching claws, and long necks that extended their browsing range.

There was no great struggle for existence in prehistoric New Zealand, and the main challenge for moa was to avoid hazardous swamps and predations by giant eagles. At the same time, the bird caused natural selection among others, for certain New Zealand plants are believed to have developed divaricated structures as a defence against the hungry moa. But after some 60 million years of the good life, the big bird was given one of the shortest and sharpest lessons in Darwinism when this country became home for another tall, flightless and hungry two-legged species.

Science in New Zealand came of age in the late 19th century, thanks largely to the contribution of Julius Haast. This explorer, geologist and museum builder owed much to the moa, from the bones retrieved from a swamp in Glenmark in 1865 to what was perhaps the climax of his glittering career – the display of skeletons at the Colonial and Indian Exhibition in London in

1886. Haast was twice knighted, first in 1875 by the Emperor of Austria – whereupon he could style himself 'von' Haast – and eleven years later by Queen Victoria – which entitled him to the letters KCMG after his name. He had taken the moa to the world, but he had also brought the world to New Zealand, for he personally gave the names of over 100 distinguished international 19th-century scientists to this country's mountains, rivers, glaciers and other geographic features. These permanent reminders of the quest for knowledge include two individuals associated with the moa: Sir Richard Owen (the 1875 m Mount Owen near Buller) and William Buckland (the Buckland Peaks and Buckland Creek in the Paparoa Range, south of Westport). Others Haast introduced to the local map were the Hookers – Sir William and son Sir Joseph – Thomas Huxley, Charles Darwin, Sir John Franklin, geologists Sir Roderick Murchison and Sir Charles Lyell, and Royal Astronomer Sir William Herschel. He also named a lake and a glacier after Ferdinand von Hochstetter, and was probably responsible for the naming of Mount Novara in memory of the ship that brought his old friend to New Zealand. Haast made a great impact on the landscape himself, for he was immortalised by the full geographic complement of a township, a glacier, two mountains, a pass, a range and a river.[60]

In the early 1860s Haast explored the mountainous regions of the South Island and became convinced that a large and unknown kiwi lived in the Alpine forests. Whilst camping out he heard the bird's call, and described it as loud enough to wake his whole party, even after a tiring day's journey.[61] The Rev. Richard Taylor had no doubts about what Haast and his companions had heard: 'the cry of that struthious giant – the moa.'[62]

# 11

## EXTINGUISHED FEATURES
### The End of the Race

*Giants of birds, the Moa, stately trod*
*By river side, and swamps of green,*
*Vanished are they, but known to God,*
*Ere man appeared to mar the scene.*

T. W. PORTER, 1913[1]

For those contemplating emigration in the mid-1800s, New Zealand was promoted as a 'promised land'.[2] According to one colonists' guide, the many attractive features of that southern land included the 'music' made by its many birds, as described by Captain Cook some seven decades earlier:

The ship lay at the distance of somewhat less than a quarter of a mile from the shore, and in the morning we were awakened by the singing of the birds; the number was incredible, and they seemed to strain their throats in emulation of each other. This wild melody was infinitely superior to any that we have ever heard of the same kind; it seemed to be like small bells most exquisitely tuned; and perhaps the distance and the water between, might be no small advantage to the sound. Upon inquiry we found that the birds here always began to sing about two hours after midnight, and continuing their music till sunrise, were, like our nightingales, silent the rest of the day.[3]

If the moa was ever part of this dawn chorus, on account of an elongated windpipe, its contribution was likely to have been more booming than melodious.[4] The enchanting music made by New Zealand's feathered tribes was probably not the inducement for many emigrants, but, even as increasing numbers of them were contemplating that journey, John Rule was transporting his small piece of moa bone to London. Had it been accepted at the Royal College of Surgeons in London it would have gone on exhibition at the Hunterian Museum, alongside the remains of other animals, either extant or extinct. Till then, large birds were represented there by the ostrich, which Richard Owen consulted when identifying Rule's specimen. However, that avian skeleton from Africa would soon be joined by its equivalent from New Zealand, as more parts came to hand.

Those obliging early moa collectors never saw their subjects alive, but New Zealand's original settlers most certainly did. However, compared to the large amount of surviving evidence, Maori traditions are surprisingly light on the subject of the moa. On the other hand, the oral record includes other strange creatures for which no remains have ever been found, an anomaly illustrated by recent events in the North Island of New Zealand.

During the October 2002 school holidays, two 11-year-old boys hunting for pheasants in the Tukituki Valley in Hawke's Bay found a 70-centimetre stick-like object lying under a metre of water at the bottom of a creek. They retrieved it and took it to their local Department of Conservation office, where the 1.35-kg 'stick' was identified as a bone from the biggest species of moa, *Dinornis giganteus*.[5] Shortly afterwards, an entirely different type of creature was reported some 300 km away in the Waikato district, and identified as the mythical one-eyed taniwha Karu Tahi, which occupied a small swampy area near the Meremere power station beside State Highway One. It was one of three mythical creatures which the Ngati Naho hapu (sub-tribe) of the local Tainui iwi (tribe) claimed lived along the Waikato River. The creature's lair lay in the way of improvements to the nation's main highway, requiring negotiations between road-builders and local Maori. Some attributed the unusually large number of fatal collisions on this section of road to the presence of the taniwha, while other citizens preferred to put their trust in modern engineering to reduce the road toll. The issue was resolved when alterations to the planned expressway ensured that the creature's swamp home would not be disturbed.[6]

Respect for the taniwha reflects traditional Maori belief in creatures which embodied the dreaded and unknown. Lizards were the visible representation of the power that caused death and disease, and so were particularly feared. Stories of such monsters are found throughout Polynesia, and early ethnologists wondered if these mostly water-dwelling creatures reflected a memory of crocodiles or other such saurian beasts encountered by the Polynesian ancestors of the Maori. A cave-drawing of a pair of taniwha in Weka Pass, North Canterbury, was interpreted by Dr H. D. Skinner, Director of the Otago Museum, as 'folk-memories of crocodiles', and he thought it may have been ultimately Indian in origin.[7] Te Rangi Hiroa – Sir Peter Buck – personally recalled that certain deep holes in rivers were 'conveniently stocked with awesome monsters termed taniwha', a form of folklore that he attributed to reducing possible juvenile drownings and thereby justifying the use of 'supernatural guards'.[8]

Whilst he was in New Zealand James Cook also learned of enormous ground-burrowing lizards which could seize and devour men, while John Liddiard Nicholas heard similar accounts of an alligator-like monster when he visited in 1814.[9] The existence of such creatures could both explain the disappearance of travellers, as well as certain landscape features. Legend has it that in the central North Island a noxious monster known as Hotu-puku was tracked down by 170 brave warriors who set rope nooses at the entrance to its cave. After killing it they found it had many features of the tuatara, on a giant scale. The heroes of this successful taniwha-slaying were then invited to the Waikato district to deal with another local hazard, a water-monster named Peke-haua. A third taniwha, named Kataore, lived in a cave near Rotorua, and it too was successfully eradicated.[10] But however unlikely, such tales of monsters and their fearless conquerors were hardly confined to the Pacific. Those colonising Britons who claimed to bring the benefits of civilisation to New Zealand also came with imaginative adventure stories of their own. St George, the patron saint and protector of England, originated in Roman times and became famous for dragon-slaying around the 13th century. (Much later, the image of this chivalrous horseman dispatching a dragon became well-known in New Zealand thanks to St George brand preserves.)

In one sense European science was quick to recognise the taniwha of New Zealand. During Mesozoic times the seas around the country were ruled by three different types of large swimming reptiles – ichthyosaurs, plesiosaurs and mosasaurs. Top of this particularly competitive food chain was the monstrous mosasaur, which grew up to 15 metres in length, and could have a metre-long jaw. Sixty-five million years ago, these beasts, along with three-quarters of all the world's living creatures, died out in a series of apparently simultaneous extinctions. Theories as to the cause include asteroids and comets, while one even suggests that the disappearance of the dinosaurs was due to chronic constipation brought on by changing vegetation.[11] Meanwhile, other convulsions to the Earth's surface included the breaking up of the giant landmass known as Gondwana, enabling New Zealand to split off and become an ideal, mammal-free home for flightless birds.

The Waipara formation in the north-east of the South Island

was known to 19th-century geologists as a rich source of marine reptiles from the Mesozoic. The first fossils were found in a ravine on one of the tributaries of the Waipara River in 1861 and sent, naturally, to Richard Owen, who published a description and presented them to the British Museum. The next discoveries here followed a great flood in 1868, but this collection – which included fragments of a plesiosaur jaw – was lost on the trip to England when the *Mataoka* disappeared without trace.[12] Fortunately, Julius Haast had arranged for descriptions and drawings to be made of the main specimens in case the unthinkable happened. He had regretted this material leaving New Zealand in the first instance, but the fact that they were destined for that 'illustrious comparative anatomist' Professor Owen was 'some slight satisfaction'. Apart from not having suitable scientific literature of his own to identify the fossils, he recognised that the Hunterian Professor had much to teach the colonial scientist and was sure he would provide 'a classical description of them, which will form the basis of reference and work for all future New Zealand Palaeontologists'.[13]

Following several more expeditions to the Waipara district, including one by James Hector who was now collecting for the recently established Colonial Museum, several tons of fossil-bearing 'cement-stone' were obtained. Alexander McKay, the skilled workman who later came into conflict with Julius Haast over Moabone Point Cave, spent three months liberating fossils from the matrix, and they were found to represent 43 marine reptiles, mostly 'of gigantic size', and belonged to at least 13 different species. To one of these long-necked sea lizards Hector gave the name *Mauisaurus haasti*, acknowledging Maui, the legendary Polynesian fisherman who discovered New Zealand. Another mass of vertebrae, skull and paddle bones was found to belong to the same species of mosasaur, and to this Hector gave the name *Taniwhasaurus oweni*.[14]

Although New Zealand's first human settlers brought

memories of similar lizard-like monsters, they were probably unprepared for the larger species of moa. Certain of these also became known by other names, such as pouakai, kuranui and te manu nui, and as such became the subjects of legends.[15] But supernatural exploits were not confined to Maori mythology, for they were matched by occasional reports of giant moa sightings in the otherwise scientifically authoritative publications of the New Zealand Institute, much to the displeasure of ethnologist Elsdon Best. To his mind, the Maori had been the source of some 'absurd statements concerning the moa', but these had been 'swamped in the flood of irresponsible nonsense fabricated and published by witless Europeans'.[16]

Albeit of otherworldly proportions, several Maori traditions of the moa make specific reference to the East Coast, a district which featured prominently in the European discovery of the bird. One story concerns Ruakapanga, who was among the first men to arrive in Aotearoa/New Zealand. Whilst hunting in the bush in the Bay of Plenty he and his companions encountered some large and unknown birds. They made a trap of plaited vines, but each time one of the birds ventured inside it managed to break free with its powerful legs. Finally, Ruakapanga made a much stronger trap, which entangled one of the birds. The warriors killed it with their spears, and it became known as Te Manu nui a Ruakapanga – the Great Bird of Ruakapanga.[17]

The same bird was also known to Pou-rangahua, who lived at Turanga, where Gisborne now stands, and had a young son who was in the habit of poking out his tongue. Pou decided that the boy was hungry and was using his tongue to point in the direction of a source of food. Determined to find it, Pou set off in his canoe, sailing north to distant Hawaiki. He made friends with people there, and when he tasted their kumara (sweet potato) he knew it was the food his son had pointed to. Pou wanted to take the kumara back to his own home, but now found his canoe was missing. The ariki (chief) Tane suggested he return to Turanga on the back of Te Manu nui a Ruakapanga, so early next morning Pou climbed on board the great bird, along with two baskets of seed kumara. They flew south, safely passing Hikurangi, the mountain home of Tama the Ogre. But when he reached Aotearoa, Pou did not descend quickly as Tane had instructed, instead forcing his steed to fly further and take him right to his home. Pou was welcomed back by his tribe, and the kumara he brought was soon enjoyed throughout Aotearoa. But because of his selfishness the great bird had been detained too long, and on its return home was caught and destroyed by Tama. This story explained both the introduction of the kumara to Aotearoa, and the disappearance of the giant bird. It was gone, leaving only fragments of bones and eggshell and its tiny cousin the kiwi.[18]

There are variations on the story of Pou-rangahua; one suggesting that he lived at the mouth of the Waipaoa River, on the western shore of Poverty Bay.[19] There was also another bird — Tawhai-tari — which had been unable to bear the weight of its passenger and baskets of kumara and so was replaced by the larger and stronger Te Manu nui a Ruakapanga. Also contrary to Tane's instructions, Pou had plucked some of the great flying bird's feathers, which became the first kahikatea tree in Aotearoa.[20] Common to all these stories is Hikurangi, a name carried from the original Polynesian homeland of Hawaiki to both Rarotonga — where it became Tama's lair — and Aotearoa. No doubt after Tama had captured Te Manu nui a Ruakapanga, large bones littered his Hikurangi mountain home, just as they would later be discovered near Hikurangi in Aotearoa.

In another story, Paoa was the ancestor of many East Coast tribes and possessed great powers. When the giant Rongo-Kako visited the district, accompanied by what was described as a huge kiwi, Paoa was determined to capture the bird and constructed a snare on the mountain Hikurangi. Rongo-Kako and his kiwi arrived from the south, leaving footprints at locations on the Mahia Peninsula and at Poverty Bay. When Rongo saw the snare he struck it with his stick, causing it to recoil and shake Hikurangi so much that pieces fell off it and formed smaller mountains nearby. The great kiwi escaped, and is said to have also left footprints on rocks beyond the East Cape. Much more down to earth were the impressions of moa footprints that were the subject of a scientific paper written in 1871 by Archdeacon W. L. Williams, son of the Rev. William Williams, the first Bishop of Waiapu. The younger Williams had become aware of these footprints some five years earlier, and the slabs preserving them were now in the collection of the Auckland Museum. Williams suggested the prints had been made by a single bird, one of the smaller species of moa, but was unsure whether they were contemporary with the numerous bones that had been collected in Poverty Bay by his father.[21]

One European who was in a very good position to gather Maori traditions relating to the moa was itinerant missionary and collector William Colenso. He was well aware of the richness of this record, noting: 'Like other rude martial unlettered nations, the New Zealanders had many traditions, legends and myths.'[22] In his first published paper on the subject – in *The Annals of Natural History* in 1844 – Colenso found the Maori 'totally ignorant of anything concerning the *Moa*', apart from a few stories he dismissed as superstition. Thirty-five years later, he updated his contribution with further details on the extinct bird, which he entitled 'What I have gleaned since'.[23] In this he stated that he was aware of only one Maori myth or legend that specifically mentioned the bird – the ancient story of Ngahue, from Hawaiki, who came to Aotearoa where he saw (and killed) the moa and also discovered greenstone.

Colenso also related the story of the ancient fire of Tamatea, which he had recently received from 'an old intelligent chief of the East Coast'. According to this, the whole land was consumed by fire and all living things were destroyed. But two moa managed to escape the flames, and survived at Te Whaiti and

Whakapunake, two places familiar to Colenso. On his first visit to the East Cape in 1838 he had heard about the 'moa' which was said to live at Whakapunake, and on his second visit in the summer of 1842–3 he investigated the mountain without success. He had since heard from another source, an 'old chief of the Ngatiporou tribe, from Tokomaru, near the East Cape', that only one moa had escaped the fire of Tamatea, and that was the elusive bird of Whakapunake.[24] Moa remains have been found in that district, and there may also be a strong basis for Tamatea's fire. There is sufficient evidence that New Zealand had been subject to extensive and mostly man-made conflagrations, with a disastrous effect on communities of bush-dwelling flightless birds. Early European explorers remarked on the frequency with which Maori set fire to surrounding vegetation in an attempt to maintain clear paths though dense forests and otherwise impenetrable tangles of fern and scrub.[25] Once out of control, fires altered the landscape and destroyed much of the forest on the drier plains and river valleys on the east of the country, which had been home to the bulk of the moa population.

Colenso was first to record the word 'moa', and found the name of the bird he described as, the 'most uncommon animal New Zealand has ever produced', something of a mystery. Colenso also understood that in the Friendly, Society and Sandwich Islands 'Moa' was the name given to the 'domestic cock' or fowl, and a word adopted by the missionaries there.[26] But the first evidence of the New Zealand 'moa' that came to light – Rule's bone and others collected by Williams and Colenso – suggested birds of unprecedented proportions. The term 'moa' therefore came to be associated with the larger species, which may not have been recognised as such by Maori. Apart from this confusion, Maori communities could not be expected to be familiar with bird populations in districts beyond their own. And as the various moa succumbed to the predations of humans, the birds most likely to survive the longest were those in more isolated regions.

By the time Europeans had arrived, the moa had long since disappeared from more accessible areas, such as in the North Island, so it was hardly surprising that local Maori had little, if any, knowledge of the bird.

William Colenso was an early collector of 'old Maori proverbs', recognising them as powerful tools when addressing the people. He amassed upwards of 1400 examples, hoping to find references to that 'almost mysterious animal the Moa'. He found only eight, whose 'very abrupt, primitive, and legendary style, and esoteric or hidden meaning' placed the moa 'very far back into the night of history!'[27] Of these proverbs, two had a strong sense of finality: *Ko te huna i te Moa!* – All have been destroyed as completely as the Moa!; and *Kua ngaro i te ngaro o te Moa!* – All have wholly disappeared, perished, just as the Moas perished; none left!'[28] Colenso also made a careful study of more than 900 pieces of Maori poetry, including Sir George Grey's published collection – 'some of them very long (and not a few of them written coarsely in a wretched hand)' – and found just five 'small scant and antiquated sentences' referring to the moa.[29]

Examining the proverbs in detail, Colenso found there were nearly 70 relating to birds alone, but just a few 'meagre and misty mythical' references to the moa. In his words, 'another pregnant omission!' And while the ancient Maori possessed charms, spells and prayers for the successful catching of such birds as kiwi, kakapo, koitareke (quail), weka, kaka, kotuku (white heron), huia, kereru (pigeon), tui, pukeko, parera (duck), whio (blue mountain duck), kawau (shag) and toroa (albatross), there was none for capturing their biggest game of all, the moa. Again Colenso was surprised, noting '[t]his alone has ever been to me an unanswerable argument.'[30] Perhaps he had already unwittingly provided an explanation for such apparent omissions, for ten years earlier he had described the Maori language as: 'remarkable for its euphony, simplicity, brevity, clearness and copiousness ... having proper names for every natural thing however small –

every part of a vegetable, for example – whether above or below ground – and upwards of fifty names for a sweet potato, and forty for a common one.' However, he found the language deficient in what he described as 'abstract ideas unknown to the New Zealanders'.[31] Perhaps, notwithstanding such tangible evidence as bones and egg-shell, the extinct moa would have qualified as an 'abstract idea' to the Maori at that time.

Other early collectors found a similar paucity of moa references. In his *Te Ika a Maui*, published in 1870, Richard Taylor included 85 proverbs, with the only one that mentioned the bird also being on Colenso's list: *Ka ngaro a moa te iwi nei* – The tribe will become extinct like the moa. And just like his colleague, Taylor discovered there were several fables relating to other plants and animals of everyday life, such as the kumara, eel, kauri, shark and rat.[32] Four years later the Rev. William James Stack published the results of his re-examination of the Maori poetry collected by Sir George Grey, and found that in nearly 500 pieces of composition the word 'moa' occurred on just seven occasions. The first of these was the most frequently cited, and in its usual version, *Ka ngaro i te ngaro a te Moa* – Lost (or hidden) as the Moa is lost. Stack offered two interpretations; that it alluded either to the Moa of Hawaiki, or the bird of New Zealand which had disappeared 'long ago'. To summarise, Stack could find no evidence in this collection of early New Zealand poetry that its composers were familiar with the moa.[33] Another who came to a similar conclusion was ethnologist Elsdon Best. He began collecting Maori lore in the 1870s, 'in the depths of the Forest of Tane, in native huts, and in military encampments, in lone places where now is heard the rushing locomotive and the whirring motor car'. Best was able to find few Maori traditions relating to the moa, so concluded the bird had 'long disappeared from the world of life'.[34]

Soldier, Magistrate and Native Land Court Judge William Gilbert Mair spent the greater part of the 1870s working to establish friendly relations with disaffected Maori following the New Zealand Wars. From this contact he wrote that: 'In all these thousands of pages of Maori lore which I have written from the mouths of [Maori] witnesses in Waikato, at Rotorua, in the Bay of Plenty, Hawke's Bay, Manawatu, Wanganui, and Taupo, there is not one word about the moa.' In 1871 he asked among Ngati Maniapoto people in the Waikato for information on the moa and was told: 'We do not know anything about it, but perhaps our ancestors did.' When Mair reminded them that they had earlier helped Ferdinand von Hochstetter dig for moa bones at various locations they claimed that they did not know 'what the creature was like until the doctor told us it was a great bird taller than a man or a horse'. Like Colenso, Mair had also heard of the strange creatures in a cave at Whakapunake, on the East Coast. A Manawatu chief told him that these, the 'last of the moas', were guarded by a taniwha ('dragon') whose 'awful loud breathing was sufficient to deter anyone from approaching'. Mair also discounted the idea that the moa was so common that the Maori 'did not take sufficient interest in it' and make it the subject of proverbs, songs or karakia (ceremonial chants). The fact was that such records existed of rats, tui and 'other small fry', which existed in far greater numbers than 'the splendid moa'.[35]

The interpretation of early Maori references to the moa was summed up in the paper read by Captain Frederick Hutton to the Philosophical Institute of Canterbury in 1891. In the large number of ancient tales and poems which had been collected and published, the allusions to the bird were in the opinion of Walter Mantell, 'very slight and obscure'. Although the Rev. Stack believed that because *Ka ngaro i te ngaro a te Moa* occurred in a very early Maori poem, the bird was not in existence when it was composed. And while alternative translations had been offered for some of the collected proverbs, Hutton cautioned that they might be no more than later deductions from the 'originals'.[36]

Hutton referred to another surviving record of the moa, in the form of Maori placenames. In the North Island Colenso had recorded some 18 different such examples, among them Moakura and Moawhango, which he translated as 'startled moa' and 'reddish brown moa' respectively. But the missionary had been unable to ascertain whether the appearance of 'moa' in a name was necessarily a direct or even contemporary reference to the living bird. In stark contrast to Colenso's findings, Hutton concluded that the South Island had no Maori placenames which contained the word 'moa'. But thanks to European naturalists, the bird had recently made its mark on the South Island. In 1852, whilst he was Commissioner of Crown Lands for Otago, Walter Mantell discovered a rich source of the bird's remains at the mouth of the Te Awakokomuka Stream near Oamaru. He named the site Awamoa, meaning 'moa stream', and it stuck.[37]

In fact as well as fiction, the moa were the most extraordinary of the New Zealand birds. They filled the roles taken in other countries by herbivores such as giraffes, rhinoceroses, and kangaroos. And while those animals had lions, hyenas and dingoes as their natural predators, all the moa had to watch out for was the giant Haast's eagle. Although there were probably not enough of these birds to cause the extinction of the moa, they had claws as big as a tiger's and, according to Richard Holdaway, could strike their prey 'with the force of a concrete block dropped from the top of an eight-storey building'.[38]

From the identification of the first bone, scientific understanding of the moa proceeded rapidly to determinations of the bird's shape and number of species. By the end of the 19th century, resistance to the idea that it had disappeared prior to the arrival of the Maori was broken down, for there was now increasing evidence that humans had everything to do with the rapid extinction of the bird. The most likely alternative cause was the bird's inability to adapt to other changes, such as variations in the New Zealand climate, but these would hardly have prompted the almost simultaneous departure of all species, which are now believed to have numbered 11.[39]

It is now generally accepted that the root cause of moa extinction arrived from central Eastern Polynesia no more than 1000 years ago. More accurate dating of this country's first settlement by humans is made difficult by the nature of surviving archaeological evidence, and the natural destruction of early occupation sites. The early settlers spread widely throughout the country, taking advantage of the large and easily accessible resources of marine mammals and flightless birds. As these became depleted, denser forest areas were systematically exploited as an alternative food supply. With growing knowledge of the seasonal habits of birds, the most vulnerable of these also became eliminated in certain localities.[40]

With no fear of humans, moa would have been both easy and plentiful prey for this country's first settlers. And although they filled several ecological niches, they could not adapt to the massive changes represented by the arrival of humans. Unlike their smaller nocturnal kiwi cousins, moa were extremely vulnerable, and there are over 300 known moa-hunting sites throughout the country, where the birds were butchered and their meat processed prior to being taken back to tribal areas. Such facilities were located at the mouth of almost every substantial river on the east coast of the South Island, and also in southern Taranaki, and some had estimated takes of between 5000 and 10,000 moa. At the largest site, at the mouth of the Waitaki River, the estimated number of moa prepared for the pot ranges from 20,000 to a staggering 90,000. The finding of egg-shell at these early meatworks suggests little thought was given to the needs of future consumers. As breeding moa were killed and replacement generations were eaten before maturity, stocks dwindled, encouraging even more intensive and devious means of hunting.[41]

To the first New Zealanders, moa stocks would have seemed

limitless. European settlers in the mid-19th century had a similar response to another natural resource, timber, for they were told that this country's extensive forests would provide 'an inexhaustible supply for the wants of many generations, both for shipbuilding and other purposes'[42]. With no concept of a limit to the supply, moa were hunted without mercy. As a rich source of protein they contributed directly to the increase in human population and – as a result – the decrease of their own. The end came quickly, for most populations of the bird were wiped out in the period between 800 and 500 years ago,[43] while man and moa may have co-existed for no more than a century in any given region.[44] With such a rapid disappearance, perhaps it was hardly surprising the bird made little impression on the oral record of the Maori. For later generations, the larger moa may have transcended the realm of earthly birds altogether to inhabit the world of spirits, much to the frustration of European inquirers who could not understand how such big 'birds' could have disappeared so completely.

The moa's fate was no isolated event, for there have been sufficient other large species' losses around the world to fuel what has been termed the 'great megafauna extinction debate'. There are two main schools of thought, with either climate change or humans to blame, with the latter now regarded as the most likely cause. To a growing number of scientists there is incriminating evidence that virtually all recent extinctions of animals worldwide occurred relatively soon after the arrival of humans in any one place.[45] In the case of New Zealand, extinctions were particularly dramatic because the country's first arrivals had come from elsewhere and were required to adjust to the demands of a new environment in a relatively short time-frame. This proved a much rockier road to sustainability than in Africa, for example, where humans and animals evolved together in relative harmony over long periods of time. But New Zealand was not alone, for what Tim Flannery describes as the 'process of environmental ruin' suffered by that country was echoed to some degree on 'virtually every Pacific island whose archaeological record has been examined'. On Easter Island, for example, the Polynesians eliminated almost the entire flora, 'one of the most striking records of forest destruction anywhere in the world', and then ate most living things off the face of the island.[46]

Further north in the Pacific, it is now believed that more than 50 species of birds became extinct in Hawaii following the arrival of humans nearly 2000 years ago. With no predatory mammals or reptiles, many birds had become flightless, and, as described by Peter Ward: 'When the first Hawaiians arrived they found flocks of striding, walking, waddling, preening, and ultimately helpless, flightless birds, all running for their lives in the face of hungry humans and dogs.' In New Zealand, the moa was hardly the only victim, for 34 bird species are known to have become extinct since the arrival of humans, and on the nearby Chatham Islands the number is upwards of 27.[47] Survival, not sustainability, was the prevailing attitude in what once appeared to be lands of plenty.

The moa are hardly alone, for some 99. 9 per cent of all species that have ever inhabited the earth have been lost to extinction. The really bad news, according to Anton Gill and Alex West, is that 'in the long term almost nothing survives'.[48] Extinction is a much safer bet than survival, and it is only because the appearance of new species has a slight edge over the disappearance of existing ones that there is any life on earth at all. It will come as no comfort to the moa, but extinction is simply evidence of evolution at work. Prior to the arrival of humans, New Zealand was dominated by an assemblage of birds, the likes of which Tim Flannery describes as 'unbelievable' and found nowhere else on the planet. This was the relatively unmodified result of a completely different line of evolution to the rest of the world, one that demonstrated what might have happened if mammals – as well as dinosaurs – had become extinct 65 million years ago and birds ruled the earth.[49] Although the moa did not receive a specific mention in Darwin's landmark *The Origin of Species* – the closest was a reference to Professor Owen's 'restorations of the extinct and gigantic birds of New Zealand'[50] – the author certainly made allowance for it with the observation that the 'endemic productions of New Zealand, for instance, are perfect one compared with another; but they are now rapidly yielding before the advancing legions of plants and animals introduced from Europe'.[51]

Archaeology now reveals how the moa met its fate at the hands of hungry humans. However, while the moa as an economic food stock died out centuries earlier, Rhys Richards suggests a European sealer in Fiordland may have had the doubtful distinction of dining on a lingering moa in 1846, while Maori in Preservation Inlet, Fiordland, may have made a meal of moa as recently as 1868.[52] We might contemplate an alternative outcome had Europeans discovered reasonable numbers of live moa when they reached New Zealand. Science would have clamoured for examples of the world's tallest bird, with Richard Owen surely demanding a complete specimen to stand alongside his ostrich in the Hunterian. If museum requirements and hungry bushmen did not push the bird to the brink of extinction, the attitude of the general public to natural resources was hardly reassuring. According to one school textbook at the end of the 19th century, young New Zealanders were encouraged to adopt a fairly calculating attitude towards their wildlife. They were told that: 'As civilised man advances and takes possession of forest and plain, the larger wild animals that formerly inhabited the region must either come into his service or be destroyed.' Most vulnerable of these were the flightless birds, for they could not escape from 'the advance of the settler, or from the keen eye of the hunter'. Unless moa flesh was tasty – the cassowary's was described as 'coarse and unsavoury' – or it produced a fashionable plumage like the ostrich, then it was of no service to man and would deservedly join the ranks of extinction, alongside the dodo.[53]

# EXTINGUISHED FEATURES

There were numerous reports of moa sightings in the 19th century,[54] but their sources, like their subjects, are long gone. The irreversible loss of the marvellous moa renders the kiwi this country's remaining ratite. It was also hunted, for its meat and feathers, but being nocturnal and ground-burrowing it managed to survive the predations of the Maori and their dogs. But more recent arrivals to New Zealand, and changes to its landscape, represent new challenges for the kiwi, and its outlook is far from assured. It would be a sad day for this country if it had to admit of its (unofficial) national symbol that it too was 'Lost as the moa is lost.'

# 12

## CREATING MONSTERS
### The Raising of the Moa

*Where moas of primordial race*
*O'erlooked the toi plumes of grace:*
*Where Tarawera's elfins played*
*Amidst the terraces they made,*
*All unforeseeing – unafraid:*
Beautiful Zealandia!

WILHELMINA SHERRIFF ELLIOT[1]

If Joel Polack's references to the remains of emu-like birds in New Zealand hadn't already raised eyebrows in London's zoological circles when published there in 1838, then Richard Owen's tabling of the bone the following year surely suggested there was something monstrous afoot. The Professor's pronouncement seemed startling enough to certain of his colleagues, but he was cautious when it came to putting dimensions on his discovery, venturing no more than it was of a 'large struthious bird' and 'nearly, if not quite, equal in size to the ostrich'. John Rule later took issue with these first suggestions of the size of the moa, pointing out that when he and Owen compared the original fragment with the largest thigh-bones of the ostrich in the

Hunterian Museum – which Owen assured him were equal to those in Paris – they found it had a larger circumference than any of them.[2]

For the time being, Owen had only a single fragment to go on. But back in New Zealand, evidence was being gathered that would soon give a better idea of the size of the bird. In his letter of 28 February 1842 which accompanied the crate of bones sent to William Buckland, William Williams suggested a maximum height at 'probably not less than fourteen or sixteen feet', and included details of claimed sightings which supported his figure. This estimate was incorporated into Owen's 1844 'Memoir on the Genus Dinornis',[3] but if this had relatively limited circulation amongst the scientific community, the concept of huge birds in New Zealand would shortly reach the more popular press. It was also given momentum by someone who should have known better – a fellow of the Royal Geological Society and founding member of the New Zealand Institute, the Rev. Richard Taylor.

In an 1850 article in a quarterly magazine, Taylor related how he had been told by a government interpreter named Edward Meurant that in 1823 at Molyneux Harbour (now Waikawa Harbour) in Southland, he had seen a moa bone that reached from the ground to four inches above his own hip, and had beef-like flesh and sinews attached. Another man who claimed to see the flesh of the moa at this time was Thomas Chaseland, described as Meurant's interpreter. A third individual, George Pauley, who was then living in Foveaux Strait, separating Southland from Stewart Island, also told Meurant that he had seen 'the Moa' near a lake 'in the interior' and described it as an 'immense monster, standing about twenty feet tall'. Incidentally, such a height would have put this bird up with the giraffe, the tallest mammal ever recorded.[4]

Such stories were guaranteed to travel, and when Taylor's got to England it was republished smartly. An 1851 issue of the *Annals of Natural History* included several matters of scientific interest passed on by an observant correspondent, the first concerning the resuscitation of frozen fish, a phenomenon that was observed during Sir John Franklin's first polar expedition. To this the writer contributed the new information that spiders 'frozen so hard as to bound from the floor like a pea' were able to be thawed and return to their normal state. He then forwarded two items on the moa, the first being a newspaper report of a recent scientific meeting in Sydney. A fossil moa bone was tabled, and members were told that Professor Owen, the 'highest authority on comparative anatomy', had pronounced the height of the bird to be 'from sixteen to twenty feet'. The tabler of this bone passed on an opinion that such items were plentiful in the South Island, large parts of which had 'never been seen by a white man', and that natives had seen such a bird alive within the last 25 years. The second moa item was an excerpt from Richard Taylor's article in the *New Zealand Magazine*, which 'accurately' repeated George Pauley's claim of a 20-foot monster but somehow managed to mangle Meurant's name to 'Memaul'.[5]

In 1855 Taylor published the first edition of his book *Te Ika a Maui*, on the natural history and inhabitants of New Zealand, which included his earlier references to giraffe-like birds in the southern regions.[6] Seven years later this tall story enjoyed a third outing, in an Auckland publication also known as the *New Zealand Magazine*.[7] But in his second (1870) edition of *Te Ika a Maui*, while Taylor did mention Meurant — now 'an old sealer' — and his horror at being offered a cooked leg of moa, suspecting it might be human, he thought better of repeating the story of the 20-foot bird. But he still perpetuated the notion of giants, including an illustration of a *Dinornis giganteus* towering over a kiwi and takahe, while an unsuspecting *Dinornis elephantopus* was about to be speared by a Maori hiding behind a ti, or cabbage tree. Taylor likened the *Dinornis giganteus* to the 'cameleopard' — an early name for the giraffe — but considered the *Dinornis elephantopus* an even 'more wonderful bird', its massive frame resembling that of the animal after which it was named.[8] One of the difficulties of dealing with moa was the lack of handy comparisons, so Taylor provided another perspective on the bird when he reported that Maori at Waingongoro had told him it was 'quite as large as a horse'.[9]

Two of the individuals party to the moa 'encounters' at Molyneux Harbour in 1823 turned up later, but unfortunately neither offered any further details on the subject. In 1842 government administrator Edward Shortland took Edward Meurant with him as his interpreter to parts of the North Island whilst dealing with Maori land issues, but there is no record of the moa being a subject of discussion.[10] Meurant was later described as an 'Australian of the early Type', and first heard of in this country as a sealer in Foveaux Strait, later moving to the Waikato district where he married a Maori woman.[11] On a tour of whaling stations in the South Island in 1843 Shortland was accompanied by Thomas Chaseland, but if the moa was not talked about on this occasion then it may have been because the latter was incapable. Shortland recorded how Chaseland had managed to get 'beastly drunk' on shore and lay like a cask at the bottom of the boat. He described him as a 'specimen of the Australian half-cast', and although an 'inveterate' drunkard was considered the best whaler in New Zealand.[12] This wild character, described as 'a giant of a man', also left his mark on the landscape, there being various parts of the southern South Island named in his honour. Most notable is Chaslands [sic] Mistake, a promontory in South Otago, but history has not recorded which of the sealer's mistakes it refers to.[13]

The story of those sightings in Molyneux Harbour would not go away, and 133 years later they were still part of some histories of the discovery of the moa. In 1956 German author Herbert Wendt wrote that the entry of 'the most popular of New Zealand animals' into the scientific world was distinguished by arguments between its claimed discoverers that were 'more bitter than any dispute fought out between authors, inventors or the creditors of a bankrupt firm'. However, the story of that discovery could be 'fairly reliably reconstructed', and in his view, both the first (oral) report on the bird and the first find of its bones were made by American sealers working in the South Island in 1823. One of these named Meirant [sic] collected 'several of the monster's bones with flesh still adhering to them', but they were lost, and the moa was forgotten. Then, in 1838 [sic] it 'caused a great stir' when a bone was brought to Professor Owen by a person Wendt described as the 'New Zealand traveller Rule'.[14]

Exaggeration and misrepresentation were hardly confined to the early sealers who came to New Zealand. The moa soon became larger than life in a wide range of publications, including a 1895 encyclopaedia, whose title – *The Universe: or the Infinitely Great and the Infinitely Little* – probably said it all. Author F. A. Pouchet claimed: 'Immensity is everywhere', from the 'azure dome of heaven' to the 'living atom too minute to display to us the marvels of its organization'. Somewhere in between lay the 'ornithological marvel' recently discovered by the 'illustrious' zoologist Professor Owen, whose part skeleton at the Hunterian Museum represented a complete bird that was 'eighteen feet high'. The bone of a man's leg was only 'a slender spindle' when compared to that colossal animal. Pouchet, the director of the Rouen Museum of Natural History, added that when first discovered, New Zealand was full of 'birds of appalling size'.[15] His encyclopaedia included the now standard image introduced by Hochstetter in 1863, of a moa towering over a trio of kiwi. It was also in *Webster's International Dictionary* at the time, this household authority defining the moa as 'any one of several very large extinct species of wingless birds', thereby ignoring all but the tallest family members.[16]

It began as a puzzle to science, but the moa soon graduated to a range of more popular and public uses, from trademarks to national identity. The late 19th century saw the recently established colony of New Zealand eager to cultivate a history of its own, being mindful of its youth in comparison with the ancient civilisation of Europe. The moa and the Maori – the one extinct and the other then perceived to be endangered – provided links with ancient times, and were seen as suitable themes. Best known in this genre was the large oil painting by Charles Goldie, in collaboration with Louis Steele, *The Arrival of the Maoris in New Zealand*.[17] Another artist dealing with big and dramatic moments in local history was Kennett Watkins, whose canvases included *The Phantom Canoe: a Legend of Lake Tarawera*, *The Explosion of the Boyd* and *The Legend of the Voyage to New Zealand*. He also turned to natural history, with *The Haunt of the Moa: a Scene in a Puriri Forest*, with the bird placed in a gloomy clearing of an ancient and primitive forest. The gnarled trunks of the giant trees are closing in as if to suggest the inevitability of extinction.[18] Poets also contributed to this notion of Maoriland, where the moa – now a 'feathered monarch' – could strut triumphant through his 'former flaxy kingdoms', and 'Maori' conveniently rhymed with 'kauri', another ancient giant.[19]

> Maoriland; my mother!
> Holds the earth so fair another?
> O, my land of the moa and Maori,
> Garlanded grand with your forests of kauri . . .[20]

In both verse and pigment the moa found itself in forests where trees assumed ecclesiastical proportions:

> Within thy dim cathedral aisles,
> More exquisite than marble piles . . .[21]

The size that had rendered the moa vulnerable to human predators also made it fair game for cartoonists. This late-19th-century lapse into levity contributed to the bird's undoing, at least as far as any serious symbolic duties were concerned. As New Zealand entered the 20th century, graduating from a colony to a dominion, it began looking to the future rather than reflecting on a mythical past. In 1899 it demonstrated its maturity and responsibility by dashing to Britain's assistance in the South African War, and from 1914 obtained further opportunities to prove its prowess on the battlefield. Those soldiers may have left home wearing the badge of the silver fern, but by the end of the so-called Great War they had forged the beginnings of a new identity. They had become associated with the bird on the tin of a popular brand of boot polish – invented in Australia, ironically enough – and so became known as 'Kiwis'. Blighted by extinction, the moa could only watch as its small and aggressive cousin overtook all other pretenders to become unofficial national symbol. Unlike the lumbering moa, the kiwi had so far eluded its predators, but it was also a bird with character, and seen to reflect the individuality of the New Zealander. Two decades later, another World War would cement the ascendance of the kiwi. The moa, on the other hand, was relegated to a subject of ridicule at the hands of caricaturists, perhaps to be turned into a beast of burden for early Maori, or the equivalent of a giant Christmas turkey.[22]

Moa were also the preserves of museums, where their skeletons – with plaster of Paris substitutes if genuine parts were missing – were reconstructed. Most of these made much of the moa's neck, stretching it unnaturally in support of its claim as the world's tallest bird. In 1899 young New Zealanders were told that the moa 'stood sixteen feet high' and was 'believed by many' to still exist in 'the unexplored regions of New Zealand'.[23] A more moderate perspective was provided a few years later in the school textbook *Nature in New Zealand*, edited by Captain F. W. Hutton, who was now Director of the Canterbury Museum. It opened with the almost obligatory reproduction of the Goldie and Steele painting, *The Arrival of the Maoris in New Zealand*, and the first chapter was devoted to the moa, 'A Monstrous Bird'. Thanks to 'unstinted thought, study and research', New Zealand could now claim as its own 'the largest, the strongest, and probably the ugliest bird on record'. The great moa now stood 'about twelve feet high', and it was a vegetarian with 'splendid digestive powers', relishing anything in the shape of roots and twigs. This voracious grazing may have been responsible for the Maori legends in which they even 'gobbled' up humans. Prior to the arrival of the Maori, the moa dominated the New Zealand landscape. It was the king among birds, and when it went walking 'all other birds quickly got out of his way'. For young New Zealanders the moa was a lesson in evolution, for it had 'paid the penalty of sloth'. When it first reached New Zealand it discovered such convivial surroundings that it multiplied rapidly, eventually growing 'fat and lazy' and losing the use of its wings. This was no great disadvantage at the time, for the bird had 'fine long legs to stride about with'. And in keeping with the growing military mood of the early 1900s, Captain Hutton suggested that a 'flock of [moa] on the march must have been a sight worth seeing'.[24]

While moa skeletons became standard features of natural history museums, some institutions went one step further. In 1912 the Auckland Museum appealed to the public and raised sufficient funds to undertake two major projects: the purchase of a set of ancient Maori carvings and the mounting of a new display of flightless birds. The latter was dominated by a restoration of the largest moa, *Dinornis maximus*, and its skeleton – a plaster-cast replica bought from a dealer – the Museum not having the resources of a local swamp to draw from. The pair stood alongside other members of the ratite family including the ostrich, emu and cassowary, the entire group enclosed in a massive plate-glass showcase. The display was completed in October 1913, and the first Sunday following its unveiling saw the Museum's greatest attendance on any one day that year – 1329. According to the Museum's annual report that year: 'As the former existence of the various species of Moa is doubtless the most important fact in the history of New Zealand fauna, it is a matter for congratulation that the Museum now contains a group which not only gives an excellent idea of the size and probable appearance of the extinct Moa, but which also clearly shows the differences between it and its nearest living allies.' In the 90 years since its grand unveiling, the reconstructed moa has had several changes of showcase and surroundings, but it remains one of Auckland Museum's most popular attractions. Visitors are invariably impressed by its size, and may also be amused to learn that this versatile bird was necessarily clad in emu feathers.[25]

To date, public understanding of the moa had been based on a range of published illustrations, reconstructed skeletons and alleged sightings, of widely varying degrees of accuracy. Greater objectivity could be brought to the subject when the contents of another swamp were revealed in the late 1930s. The moa had long been associated with antiquity, so it was fitting that it should now forge a link with ancient Egypt. Barely 15 km – as Haast's eagle (*Harpagornis moorei*) might once have flown – from Glenmark in North Canterbury, lay Pyramid Valley. Here, in a depression amongst limestone outcrops eroded by wind into pyramidal forms, lay a tussock-covered swamp. Arnold and Joseph Hodgen had farmed this land since 1924, and allowed horses to browse on the tussock. One day in 1937 Joseph and his son Rob chose a cleared spot on the swamp to bury a dray horse, and had not finished digging the grave before three massive leg-bones were unearthed. Recognising them as moa, they assumed the Canterbury Museum had no need of any more, knowing that it already had the world's largest collection of such items. The three bones were put aside in the woolshed, which often doubled as the farmer's museum. It was here they were spotted in 1938 by an amateur naturalist who knew they would most certainly be of interest to the Canterbury Museum.[26]

In February 1939 a seven-man team from the Museum arrived at Pyramid Valley. They were ready to exhume the dray horse skeleton and recover any other moa bones that lay beneath it, when probings with a gum spear indicated that deposits of bones existed right throughout the three-acre site, and to a depth of 12 feet. Shallow pits were dug, all of them revealing moa bones. In expectation of Canterbury's greatest moa rush, excavations got under way, and continued until 1941 when petrol and staff shortages caused by the Second World War made further progress impossible. It was another eight years before things returned to normal, and the first systematic excavations began in 1949. Joseph Hodgen's woolshed was now used as storage for much larger quantities of bones, and before long 60 filled boxes were carried back to the Museum at Rolleston Avenue for study. The first almost complete skeleton assembled from these remains was displayed at the Canterbury A & P (Agricultural & Pastoral) Show in 1949, but the Museum itself had to be enlarged before it could put all its new skeletons from Pyramid Valley on permanent exhibition.[27]

It was estimated that moa had met their sticky end at Pyramid Valley at the rate of some 800 to the acre.[28] This moa graveyard also proved rich in the undigested remains of the birds' last meals, and the gizzard stones needed for grinding seeds, twigs and coarse grasses that constituted their diet. The average gizzard stone found with the *Dinornis maximus* was about the size of a 'half-crown' – today's 20-cent piece – although some could measure up to four inches (100 mm). The average bird contained 220 such stones, and along with food remains these now filled a seven-pound biscuit tin.[29]

Because it was excavated scientifically and was the first to give up 'more or less complete skeletons', Pyramid Valley was probably the most important moa swamp. Its holdings enabled scientists to clear up confusion over the classification of certain species, and even restored the correct head to at least one. Findings now put paid to the idea that all moa were giants, for only *Dinornis maximus* exceeded – admittedly by a considerable amount – the height of a tall man. *Pachyornis* may have been his equal, but that tall human would have certainly looked down on *Euryapteryx* and *Emeus crassus*. And while *Dinornis maximus* may have had graceful proportions like those of the ostrich, *Pachyornis* and *Euryapteryx* had 'short, bandy, ridiculously massive legs' which, in another rural analogy, were likened to a 'forty-gallon drum supported on knee length gum boots'.[30] Put more elegantly, Canterbury Museum director Roger Duff said that *Dinornis maximus* once 'stood higher than any living mammal except the giraffe and African elephant', and probably required as much grass per day as a bullock.[31]

Resurrected mostly from swamps, moa went on to dominate natural history galleries in museums around the world. New Zealanders contemplating such reconstructions might wish there was still some chance of encountering a live bird in its natural habitat. Perhaps the reappearance of the takahe in 1948 brought new hope, for Geoffrey Orbell was reported as saying he had no doubt that the moa 'still existed at the beginning of the century', although not in the Murchison Range where he rediscovered the takahe.[32] But claimed moa 'sightings' themselves almost became extinct in the 20th century, the most notable probably being that of a 'two-metre high' specimen in a riverbed in the Cragieburn Range in Central Canterbury in 1993. The fact that it was made by a publican might have been good grounds for suspicion. While various authorities would have welcomed a genuine sighting, on this occasion they didn't fancy the chances – the Department of Conservation putting them at a 'billion to one'. But it was pointed out that New Zealanders live in a very empty land, and that there might still be room left for 'the last moa'. If scientists were disappointed by this particular 'sighting', the moa was still good for a laugh, and advertisers and cartoonists had a ball, with the manufacturers of the classic New Zealand-made

Swanndri shirt proudly pointing out that this 'moa spotter' had worn one of their products.[33]

With the survival of the moa no longer an issue, New Zealanders can at least debate the height of their lost birds. Following analysis of Pyramid Valley finds there was still lack of agreement on the way the moa carried its neck, whether upright like a crane or low and looped like a swan. Until this was resolved, the height of fully-grown adult moa could only be given within a broad range: from 10 to 12 feet in the case of *Dinornis maximus*.[34] Nearly four decades later the moa's neck was still a problem, and even once supposedly early North Otago rock drawings showing moa with the neck held upright were not considered reliable. These depictions may have followed the introduction of emus, or if they were of actual moa, may have shown them in an 'alarm' posture with necks erect.[35] By the late 1980s there were increasing suggestions that the 'ostrich-type' pose given to moa exhibits in museums was incorrect. One who responded to this charge, and the suggestion that museums had knowingly exaggerated the height of their moa in order to increase the spectacle, was Auckland Museum ornithologist Dr Brian Gill. He pointed out that no less a scientist than Richard Owen had concluded that in 'the ordinary upright posture' the largest moa reached 3.1 m tall. The Museum's fully-feathered reconstruction of the giant moa, *Dinornis giganteus*, stands resolutely at just over 3 m, and was based on Owen's work.[36]

But a decade later, the moa had finally been brought down to earth, at least by Richard Holdaway and Trevor Worthy. They viewed many museum moa as the work of over-enthusiastic zoologists, producing 'anatomically impossible' reconstructions. From a reassessment of the size and relative placement of moa vertebrae they now determined that the anatomy of the living moa was more horizontal than vertical, and the birds carried their necks in a snake-like descending-ascending loop. In normal circumstances the head was at a lower point than the top of the back, the bird's highest point, which in the case of *Dinornis giganteus* was (only!) about two metres. When reaching upwards for food the moa obviously increased in height, but only then may have matched that of the large ostrich.[37]

The public imagination can now be exercised by the prospect of science being able to genetically engineer a moa revival, but any such return to the past is probably as unlikely as it is undesirable. The moa, that 'most important fact in the history of New Zealand fauna' was first identified on the basis of a 15-cm length of bone. But 140 years later, with much less evidence, this country learned it had once been home to even larger terrestrial creatures. In 1979, after nearly eight years of searching a late Cretaceous (87–65 million years old) fossil site in northern Hawke's Bay, amateur palaeontologist Joan Wiffen, of Haumoana, near Hastings, found a single 8-cm fragment of bone. It was identified as a vertebra from the tail of a small carnivorous dinosaur, about 4 metres long and weighing 0.4 tonnes. Whereas the first historic moa bone was necessarily consigned to Richard Owen in London, it was a measure of scientific progress in this part of the world that New Zealand's first dinosaur fragment now only needed to travel across the Tasman to Australia, for the attention of palaeontologist Dr Ralph Molnar at the Queensland Museum, to receive positive identification.[38]

# 13

## BACK TO THE BONE
## Natural History Museum Specimen No. 44639

> *I abide in a goodly Museum*
> *Frequented by sages profound,*
> *In a kind of strange mausoleum,*
> *Where the beasts that have vanished abound,*
> *There's a bird of the Ages Triassic*
> *With his antediluvian beak,*
> *And many a reptile Jurassic,*
> *And many a monster antique!*
> 'BALLAD OF THE ICHTHYOSAURUS', 1885[1]

Richard Owen's long life (1804–92) stretched across most of the 19th century, and for nearly half of this he was concerned with the moa. The extinct bird from New Zealand was the subject of the longest series of single-genus papers of his career, extending from 1839 to 1879. In fairness, however, it should be recorded that Australian mammals represented his longest-running theme, beginning in 1832 and ending in 1888 when, as Nicolaas Rupke puts it, 'a final paper on the subject flowed from his trembling hand'.[2]

From his appointment as Assistant Conservator at the

Hunterian Museum in 1827 until his retirement from the British Museum at the end of 1883, Owen devoted 57 years to the study of scientific specimens of national importance. His life coincided with what has been termed the age of museums, for some 200 such institutions – metropolitan, provincial and university – were founded in Britain during the course of his career.[3] Four years before Owen took up his appointment at the Hunterian, work began on the present building of the best-known museum of all, at Bloomsbury, and continued for the next 20 years. Richard Owen would play a major part in later developments at the British Museum, but in the meantime he ensured that its displays included an extinct bird from New Zealand.

Visitors to the British Museum in the late 1840s included young Frank and his sister Lucy. They had planned a visit to the natural history displays, accompanied by their Aunt Edith, and their anticipation was heightened by the remark of a friend who had never seen 'so much of God before as was there brought to view'. Frank and Lucy were no strangers to such places themselves, for they enjoyed collecting and arranging 'curiosities' – the sort of things others might regard as 'rubbish' but which they termed 'a museum'. This wholesome pastime had the full support of their father, who was determined they would not lapse into 'desultory habits'.[4]

It may have been entirely fictional, but Frank and Lucy's big day out was the basis of a cheerful guidebook, *More Pleasant Mornings at the British Museum*. Published by the Religious Tract Society, it was hardly surprising that it was subtitled 'The Handy-Work of Creation'. By observing and asking questions of knowledgeable Aunt Edith, the young pair took the reader on an instructional tour. It began, at their insistence, in the Museum's mineral section, where they saw evidence of 'omnipotence manifest in the mineral riches of the Earth'. They learned that modern science had now identified some 51 different elements – a huge advance on the six mentioned in the original chemistry manual, the Books of Moses. Frank and Lucy wandered wide-eyed amongst skeletons of the megatherium and mastodon, surrounded by glass-topped cases and shelves of antlers, tusks and bones that reached to the ceiling. They witnessed the 'boundless riches' provided by the Great Creator, and were told how learned men studied fossil remains to understand the beasts of ages past. They read about New Zealand, marvelling at a 'perfect' 5 feet 6 inch high skeleton of the *Dinornis elephantopus* and leg-bones of *Dinornis giganteus*, which suggested a bird reaching nine feet, and studied an elongated and un-kiwi-like specimen of the *Apteryx*. When Lucy drew attention to the remains of the dodo from Madagascar, Aunt Edith explained that Professor Owen had been working on fossil bones from that same island which might throw further light on the subject.[5]

By the time Frank and Lucy's exploits had appeared in print, Owen had been working on a lot more than dodo bones. He had named the dinosaurs, and hoped they might extinguish some of the new and progressive ideas about the origins and development of life. He had also identified fossils collected in South America by the young naturalist Charles Darwin who, ironically, was gathering certain information for a theory of his own. Among such bones were those of the megatherium, which went on display at the Hunterian Museum. Later, authorities there had nervously greeted the arrival from Buenos Aires of the entire fossil skeleton of a related mylodon, a giant sloth-like creature. This consignment had cost the College 300 guineas and arrived in a dozen boxes, but when opened was found to consist of 'seemingly unintelligible crumbling fragments'. However, under the 'clear-sighted comparative anatomist every bone took its proper place, and the whole fabric – as if under a wand of enchantment – rose like an exhalation'.[6]

Beyond the correct assembly of its skeleton, this mylodon provided a further test of the inductive powers of the scientist, for it bore signs of fractures at the back of the skull. Owen deduced that if such injuries had been inflicted by a carnivore, the stunned mylodon would have been easy meat. In which case, if its skull survived in a fossilised state, the broken bones would not have exhibited any evidence of 'reparative processes' as was seen here. There was no evidence that the blows, which had temporarily disabled the mylodon, were caused by another animal, such as an 'irate megatherium'. Owen therefore explained the injuries as the effects of an inanimate force, and consistent with the mylodon's habit of uprooting large trees, and the occupational hazard of being struck by falling trunks and branches. In the case of the Hunterian's mylodon, the stunned beast's fractures – one healed and the other partly so – had been fossilised for posterity. Prehistoric scar tissue thus became more grist for the comparative anatomist's mill.[7]

While the Franks and Lucys of the world studied the fruits of Richard Owen's labours, the Professor himself was, as usual, busying himself on several fronts. His single largest responsibility was the compilation of the catalogue of John Hunter's specimens, but this was frequently held up when other duties called. Such a diversion occurred in 1838 when Owen dissected and described the anatomy of a recently purchased rhinoceros. Three years later, at least the catalogue of the physiological specimens – 4404 in all – could be ticked off. As the long-suffering trustees noted, Hunter had died in 1783 and it was 'high time for completion of this task'.[8] In 1849 Owen took more time out from the catalogue to study the effect of castration on the antlers of fallow deer from the park of one of the trustees.[9] In 1857 it could be reported that every specimen of the Hunterian Collection was now on display and, finally, the long-awaited catalogue was completed in early 1860, 'a relief after so long and tedious a delay'.[10]

Under Richard Owen's guidance the Hunterian was transformed, and even by 1840 was described as 'the most beautifully planned and the most conveniently arranged museum perhaps in Europe'.[11] That compliment was paid by William Swainson, although not the one who was Owen's friend and came to New Zealand in 1841 as Attorney-General. This William Swainson was a naturalist and writer who had entertained hopes of gaining a post at the British Museum. Unsuccessful in his attempts, he too came to New Zealand, also arriving in 1841, where he took up farming near Wellington, remaining there apart from a period in Australia when he studied botany at the request of the state governments.[12]

Richard Owen was 'the personification of comparative anatomy throughout the first half of the nineteenth century'.[13] By 1856 he had 'reached the zenith of his fame'; recognised throughout Europe, he was an honorary or corresponding member of nearly every important scientific society in the world. At the age of 41 he was offered a knighthood by Sir Robert Peel – which he declined – although he was eventually made a KCB in 1884. In 1851 Owen received both the Prussian Order 'Pour le Mérite' and the Royal Society's prestigious Copley Medal.[14] The following year he was notified that he had been elected a Member of the New Zealand Society – no doubt for services to the moa.[15]

However, if the Hunterian was orderly and attractive, it was a very different story at the British Museum. That institution now claimed the largest zoological collection in Europe, and one that was growing at an alarming rate. Working conditions were far from ideal, particularly when the London fog penetrated the storage areas and staff could barely see the specimens in their care.[16] Offices were 'dark, dirty and overcrowded', salaries were small and pensions nonexistent, while breakdowns and even suicide among the staff were not unknown. In 1856 the natural history section came under the control of Principal Librarian Antonio Panizzi, to the alarm of many. The Museum now desperately needed a strong advocate for science, and Owen was the obvious choice, having transformed the Hunterian Museum into a great centre of comparative anatomy that rivalled that at Bloomsbury in some areas. In need of a fresh challenge, he accepted appointment to the newly created post of Superintendent of the natural history department at the British Museum, responsible for botany, geology, mineralogy and zoology. But his new position was not altogether welcomed by his department heads, who resented the imposition of another level of management. As a result Owen became thwarted by bureaucracy, and his duties were largely nominal. But it was a situation he turned to his advantage by simply directing his considerable energies elsewhere. There were new and magnificent collections at his disposal, providing more raw material for his unending publication programme – which continued to involve the moa. But there was also a much bigger issue: the need for a new and separate museum.[17]

Concerned over the inadequate amount of space devoted to natural history at the British Museum in its Bloomsbury building, in 1859 Owen submitted a strong report to the trustees. The time had come to split the old British Museum in two: dividing the works of Man (books, manuscripts and antiquities) from those of God (natural history). He offered his vision for a new national institution dedicated to science, and one worthy of the British Empire and its place in the world. The idea of splitting the Museum was accepted, and led to the decision to build in a new location. Land was bought at South Kensington in 1863, although construction would not begin there for another decade. The new site was not without its critics though, with one claiming it was too distant for working-class visitors, and 'might as well be in New Zealand as South Kensington'.[18]

Owen advocated that the galleries in the new museum should reflect the natural world itself, and that as many specimens from every known animal, vegetable and mineral should be shown for the benefit of the public, believing a careful study of these would lead to a better understanding of the Creator Himself. Owen's ambitious scheme – which he termed an 'Index Museum' – demanded a large amount of space, and would consist almost entirely of public galleries. It would also reflect Imperial Britain's conquests abroad, and include such ancient monsters as the mastodon, megatherium and moa. But most other naturalists were opposed to Owen's views, believing the public only needed to see a carefully selected sample of the riches of the natural world. They feared visitor fatigue and maintained that the vast majority of specimens should be kept in storage for the benefit of specialists and students. Aside from the museological debates, however, Britain had more pressing issues on its mind at the time, among them the expense of the war in the Crimea. But the ambitious Owen persisted in his dream, and had a strong ally in politician William Gladstone. His museum did eventuate, albeit not as he had envisaged it, for in the meantime the world had moved on, and many of the new developments had been fuelled by Owen himself.[19]

By 1843 Owen had already produced as many as 250 separate scientific memoirs.[20] Nine years later it was claimed that a catalogue of his works suggested the 'labours of an octogenarian', for although he was not yet 40 years old his writings 'form[ed] indeed a library'. The *Transactions* of various scientific societies teemed with Owen's papers, demonstrating his 'easy command of vast learning, his splendid felicity of illustration – the results of the most patient and accurate investigation, and of the deepest thought'.[21] But, as the *Dictionary of National Biography* pointed out, 'no account of Owen's enormous contributions to scientific literature would be complete' without mention of his habit of recycling, by repackaging – with slight alterations and additions – and privately producing copies of papers previously published elsewhere. These were later issued and sold as independent works, with little to suggest they had appeared earlier. The *Dictionary* specifically mentioned his 1879 *Memoir on the Extinct Wingless Birds of New Zealand, with an Appendix on those of England, Australia, Newfoundland, Mauritius, and Rodriguez* as an example of this 'confusing practice'.[22]

Scientists could be a devious breed, for certain early 19th century writers took advantage of scientific societies by having the free copies of their memoirs (and others bought at a nominal rate) bound together with a new title page and sold to their personal advantage.[23] Even Owen's first public use of the term dinosaur – in 1842 – had its complications. There was a smaller run of this publication dated the previous year, which may simply have been a mistake. However, as Stephen Jay Gould suggests, it could have been a sly means of backdating, and therefore claiming personal priority in the competitive business of taxonomy and etymology.[24] If Owen's habit of double publication created confusion for others, he could also be confused himself. A letter from Herbert Rix, Assistant Secretary of the Royal Society at Burlington House advised 'Dear Sir Richard' that 'you cannot have series B both in volume form and in separate papers. You can have whichever you like but only the one.'[25]

That said, Richard Owen was a prodigious writer who also enjoyed the work of others. He was a great reader of poetry and romance, and could repeat by heart pages of Milton and his favourite authors. He also had a passion for music, was no mean vocalist and was also able to play the violoncello and flute. Neither did he allow his burgeoning scientific responsibilities to prevent his enjoyment of public concerts, and is reported to have seen Weber's romantic opera *Oberon* 30 nights in succession when it opened in London.[26] Perhaps it was Weber's music that kept drawing him back, or he may have also identified with the stridently martial lines from the libretto by J. R. Planché:

> Yes, even love to fame must yield,
> No recreant knight am I;
> My home it is the battlefield,
> My song, the battle cry!
> Oh! 'tis a glorious sight to see!

The charge of the Christian chivalry,
When thundering over the ground they go,
Their lances levell'd in long, long row!

With his phenomenal powers of memory, it was little wonder that Owen had 'an unfailing flow of anecdote'. It was also said that he had a great sense of humour and a strongly developed imagination, and could be a delightful companion. Returning to his attraction for *Oberon*, Owen's personal battlefields were the meetings and publications of London's learned societies, but this knight (at least from 1884) was not always either Christian or chivalrous. He knew how to thrust his lance, for as the *Dictionary of National Biography* put it, 'no man could say harder things of an adversary or rival'.[27]

With Owen's appointment to the British Museum, that institution now had two empire-builders on staff. The first of these was Panizzi, who had been appointed Keeper of Printed Books in 1837 and through an active policy of purchasing and taking advantage of the Copyright Acts – which required publishers to deposit copies of all their books in the Museum – had nearly doubled his stock of books in ten years. When elevated to Principal Librarian, and therefore Director, he promoted a scheme to accommodate his growing collection. Architect Sydney Smirke designed him an innovative drum-shaped and dome-roofed Reading Room, which opened in the empty two-acre courtyard at the heart of the building in 1857.[28] Due to the continuing need for storage, the remainder of this central space soon became occupied by a succession of makeshift buildings. In 2001, in anticipation of the British Museum's 250th anniversary, Panizzi's now famous round Reading Room and the 'lost space' that surrounded it was reclaimed and restored as part of the Great Court project.

Owen himself had grand plans for natural history, that would both reflect Britain's standing in the world and assist the public's appreciation of the works of God. But there was now a rising generation of young scientists who were determined to rattle the foundations of this orderly framework. This network of individuals, including Thomas Huxley, John Tyndall, Charles Darwin and Joseph Hooker, wanted to professionalise the relatively new pursuit of science and stamp out the old amateur approach that prevailed in the learned bodies. They aimed to wrest it from the grip of the Anglican old-boy network, based on Oxford and Cambridge, which saw science as providing a useful proof of God's design. All His power, wisdom and goodness were to be seen in the obvious perfection of the natural world, but the young upstarts did not want their science put at the service of theology. They resented the comfortable sinecures and privileges enjoyed by clerical naturalists, whom they dismissed as 'spider-stuffers'. Richard Owen – who Huxley described as a 'queer fish' – sat at the apex of a hierarchy of Oxbridge divines and politicians, and represented all the younger scientists opposed, while another powerful member of the establishment was combined cleric and geologist William Buckland.[29]

At Oxford and Cambridge in the first half of the 19th century, teaching had been geared to the education of the future clergy. From 1824, undergraduates at Trinity College, Cambridge were required to attend morning chapel at least five times every week, and the same number of times in the evening. As far as the authorities were concerned, attendance was seen as not so much a duty as a privilege, and there were penalties for absenteeism, while habitual transgressors could be removed from the College altogether. But attitudes were changing, and when students noticed their masters' places in chapel were regularly vacant a 'Society for the Prevention of Cruelty to Undergraduates' was formed. Members produced marking sheets showing the attendance of the fellows in chapel, posting them on College notice-boards and even sending them to London clubs. In 1841 chapel rules were relaxed, so the SPCU had served its purpose, but it cheekily awarded a prize for most regular attendance. It went to Charles Perry, who was obviously a worthy winner for he went on to become the first Bishop of Melbourne, Australia, in 1847 and was described as a 'stout evangelical churchman, equally opposed to ritualistic and rationalist tendencies'.[30]

While God was making slightly fewer demands on undergraduates' time, He was also about to have some of his workings explained in a controversial book. *Vestiges of the Natural History of Creation* was published in 1844, its author (Scottish publisher Robert Chambers) wisely remaining anonymous because its subject of transmutation was a touchy subject among Victorians. The book suggested life had begun as an electrochemical reaction, and under God's guidance had developed from a simple organism to humans. This best-seller created a more sympathetic audience for the ideas that Darwin would unleash in 1859 with *The Origin of Species*. Owen was strongly opposed to such views, attacking Darwin's book in an acerbic and anonymous review in the *Edinburgh Review* in 1860.[31]

But Owen's opinions on these matters could be difficult to determine, for he did believe in a mild form of evolution, albeit one reliant on divine intervention. This was apparent in what he termed 'homology', as when the same organ in different animals was subject to different forms and functions – the bat's wing, seal's flipper and human hand, for example. He saw these as having a common structure, leading to the idea of a structural scheme for all vertebrates which he called the 'archetype'. It was by means of this device that the Creator enabled each species to be perfectly adapted to its particular lifestyle. Owen spoke of the 'continuous operation of creative power' and believed this enabled animal species to appear in a 'successive and continuous' fashion, each in a jump from its antecedent, and not by one gradually transmuting into the next. Darwin did not believe in the 'continuous operation' of a providential God, which would, of course, be equivalent to a continuous miracle.[32]

By the mid-1840s the English public had been introduced to the moa and the dinosaur, and a decade later a live gorilla went on exhibition for the first time. There was much talk about man's alleged common ancestry with such creatures and Richard Owen was determined to stamp it out, considering it treacherous and a threat to humanity's high-born status in creation. Thomas Huxley, known as 'Darwin's Bulldog' for the staunch support he lent his colleague, never missed an opportunity to bait Owen, most famously on the subject of the comparative anatomy of apes and man. If the public felt a little uneasy about their ancestry, at least the debate had its lighter moments, and cartoonists had a field day. Another leap of faith occurred in the 1860s with the discovery of a fossil Neanderthal skull and stone tools alongside extinct animals. This appeared to be pushing human origins far beyond 4004BC, which had been calculated according to Old Testament genealogies and was boldly pronounced by many 19th century Bibles as the year of the Creation.[33]

Richard Owen had the unusual distinction of being mentioned in *The Origin of Species* for both his assistance and resistance. Darwin began the 4th edition (1866) of his book with a review of opinion on the subject, and referred to Owen as 'this eminent philosopher' who had noted that the *Apteryx* and the red grouse had first appeared in New Zealand and England respectively by some unknown process. Darwin noted that since the first appearance of his book, Owen had 'clearly expressed his belief that species have not been separately created, and are not immutable productions'. Owen had also accepted a degree of natural selection, and Darwin was surprised that such an admission had not been made earlier, referring to a claim by Owen that he had promulgated the theory of natural selection as long ago as 1850.[34] This, ventured Darwin, 'will surprise all those who are acquainted with . . . his works . . . published since the *Origin*, in which he strenuously opposes the theory.' Darwin was hopeful that Owen's opposition would now cease.[35]

While he may have discovered a mechanism for explaining the development of living organisms, Darwin was quite unable to fathom the convoluted writings of Richard Owen. By at least the 6th edition of his work it was 'consolatory' to him that others also found 'Professor Owen's controversial writings as difficult to understand and to reconcile with each other' as he did. But ever generous, Darwin concluded the matter by suggesting it was 'quite immaterial' whether or not Owen preceded him, for as he had already shown in his historical sketch, both of them had long been preceded by others.[36]

New Zealand managed a few modest references in *The Origin of Species*. Darwin referred to Owen's restorations of that country's 'extinct and gigantic birds' which were allied to living species – presumably meaning the kiwi, for example – to illustrate the 'law of succession types' and the 'wonderful relationship in the same continent between the dead and the living'.[37] On the subject of birds' wings, Darwin noted that those of the loggerheaded duck were used solely as 'flappers', the penguin's functioned as fins in the water and front legs on land, the ostrich's were sails, while those of the *Apteryx* had no functional purpose at all.[38]

While the evolution debate raged, Owen proceeded with his plans for the new museum at South Kensington. The chosen architect, Alfred Waterhouse, also believed the building should be a tribute to God and the wonders of nature, and decided on a Romanesque style. The museum finally opened in 1881 and was described, among other things, as 'A true Temple of Nature'. However, some considered it too elaborate for its own good, the decoration threatening to overwhelm the exhibits. It had been designed to unashamedly devote its contents to the glory of God, but by the time it opened much had changed. Natural history was now a science, and separated from theology and the church. At the same time, Owen's 'Index Museum' concept had been devolved, with only key specimens finally going on public display. Two years after the grand opening, Owen retired as superintendent and was replaced by William Flower, who became the first director. He was a confirmed evolutionist and turned the main hall into a display illustrating natural selection, reflecting the influence Darwin's theory now held on scientific thinking.[39]

At the annual conference of the Geologists' Association in 1873, the President Henry Woodward offered a tactful summary of scientific progress during the previous year. Although scientists were now pushing back the age of the Earth from a few thousand to several millions of years, it was increasingly a case of *tempus fugit*. According to the President: 'Time, upon whose ample store the geologist is wont to draw with an unsparing hand, seems to have accelerated his pace in the nineteenth century, and hurries onward as if his few remaining sands were well-nigh run.'[40] Woodward likened the present accumulation of knowledge to 'some mighty Atoll in the Pacific', an image that would have appealed to Darwin, whose own theory on such geological formations was included in *The Voyage of the Beagle*. Darwin suggested that, like a geologist who had 'lived his ten thousand years' − a further denial of a 4004BC origin − and 'kept a record of the passing changes', we would learn much about the Earth by studying the atoll.[41] Woodward then turned to that 'intellectual Atoll' to consider recent discoveries in geology and palaeontology. The glamour subject was now the meteorite, for such arrivals from outer space demonstrated the continuity of matter throughout the universe, and contained 19 of the now 64 known elements.

In Woodward's review of palaeontological progress, Richard Owen still figured prominently. He had recently described the remains of several species of Australian wombat, two of which were very large and extinct. Now speaking like an advocate for the idea of the 'survival of the fittest', Owen was quoted as likening the latter pair to the gigantic wingless birds of New Zealand, whose size and bulk appear to have been a disadvantage in the 'contest for existence'. In contrast, the moa's distant cousins, the burrowing kiwi and the smaller wombats, had survived. Owen still cast a long shadow over geology, for Woodward referred to the Professor's achievement of 1839, 34 years earlier. Since then, various other naturalists, collectors and colonists had sent 'home' so many remains of the moa that 17 species (or, as Woodward qualified, *varieties*) had been described by that same 'indefatigable anatomist'.

Turning to the bigger picture, the President of the Geologists' Association believed that the majority of botanists and zoologists now accepted the 'Doctrine of Descent with Modification'. Although some still found such ideas 'repugnant', he respected their opinions for the controversy generated was likely to dispose of 'worthless notions'. In his opinion Darwin's theory had already passed the test, and, speaking as a true geologist, he suggested that, 'like crude ore it has been washed, sifted, crushed, roasted, and smelted, and at the end the pure metal remains.' Owen's name came up again when Woodward described him as a 'most advanced Evolutionist' who believed that all forms of both vertebrate and invertebrate life were due to 'Secondary Cause or Law', and not to natural selection. Secondary causes explained operations (as opposed to primary causes which dealt with origins), and, as for Owen's beliefs, the President could only echo Darwin: 'Upon the nature of these very delicate and baffling distinctions I feel myself quite unable to enter on the present occasion.'

Darwinism prevailed, but 27 years after *The Origin of Species* had appeared, disbelievers still turned to Owen for support. In 1886 a correspondent asked if the Professor could kindly advise of any French or other foreign writers who had 'written ably against the dogmas of the Darwinian School' – 'One or two books or pamphlets would suffice.' [42] With the growing acceptance of the new science, Owen's profile was reduced, and he retired at the end of 1883 at the age of 80. He died of old age in 1892, outliving Darwin by a decade. At a British Museum dinner in his honour, a commemoration card recorded some of the many animals that had come under Owen's scientific scrutiny. A gorilla up a palm tree cast an eye over a heterogeneous menagerie that included a nautilus, kiwi, platypus, Irish elk, dodo and mylodon, and, most prominent of all, the skeleton of a moa.[43]

Some 120 years later, Richard Owen maintains a formidable presence at his old stamping grounds. Beyond the Royal College of Surgeons' facade of six Ionic columns, a staircase leading up to the Hunterian Museum is overlooked by portraits and busts of distinguished past presidents, in red gowns and chains of office, and holding their pipes and books of anatomy. At the entrance to the Museum itself is a portrait of John Hunter, pondering a skull. Exhibits inside testify to his imagination and breadth of inquiry, and his indefatigable efforts to understand the workings of the body. They include the stomach of a gull, which Hunter kept for a year, living 'contrary to its nature, upon grain', and whose posthumous dissection showed the muscles had increased, thereby supporting his theory of the dependence of structure on function. He was also interested in suspended animation, and considered asking a man to give up the last 10 years of his life, to be frozen for a thousand years, and then thawed out and refrozen over successive millennia. The label does not mention whether Hunter planned to follow suit in order to monitor his experiment. But if he didn't have any human volunteers, he did experiment – unsuccessfully – by freezing the ear of a rabbit.

The Hunterian retains a few melancholy reminders of the beasts it once housed, among them the charred skull of a hippopotamus, a victim of the bombing in 1941 in which the building caught fire and a falling girder pierced the bricked-in basement where specimens were stored, and about one-third of the collection was destroyed.[44] During the previous 1914–18 War, the question of danger from 'aerial craft' had been considered and it was left to the discretion of the Conservator to remove Hunterian specimens to safety if necessary. Bombs were dropped from Zeppelins, but all fell at a safe distance from the College.[45] Another historic loss was Owen's 'little room upstairs' where, according to Adrian Desmond: 'The world's biggest – dinosaurs and extinct New Zealand moas – were born.'[46] German bombing laid waste to large parts of neighbouring Holborn to the east, probably including another address that played an important part in the identification of the moa – 66 Fetter Lane. The premises from whence Dr John Rule set out for his historic meeting with Richard Owen have long gone, and are now occupied by a Sainsbury's Supermarket and Business Centre.

Following postwar rebuilding, on its 150th anniversary the new Hunterian Museum was officially opened by the Prime Minister Harold Macmillan. Representing the Royal College of Surgeons at the opening was its President, Sir Arthur Porritt, who in 1967 became the 11th Governor-General of New Zealand, the first to be born in that country.[47] Richard Owen's profile is maintained by a brooding larger-than-life bronze bust by Sir Alfred Gilbert RA (1854–1934). Shrouded in his voluminous gown, the Professor examines a small specimen with a hand-held magnifying glass. Appropriately, nearby stands the articulated skeleton of a moa. This *Anomalopteryx didiformis*, presented by Dr Julius von Haast in 1873, keeps company with other extinctions in the bird world, the dodo, great auk, and the solitaire from Rodriguez in the Indian Ocean. And while William Buckland and Gideon Mantell have both been represented here by their bodily parts, it seems the Hunterian Professor made no provision for himself to be immortalised in this manner.

In 1963 the British Museum (Natural History), as it had been known, became fully independent of the original institution in Bloomsbury and was renamed The Natural History Museum. Richard Owen is well represented here too, his likenesses including an 1844 painting by Henry William Pickersgill. In his left hand the Hunterian Professor holds a moa bone, the largest tibia that came with the consignment sent by William Williams to William Buckland. Sir Robert Peel prevailed upon William Buckland to ask Owen to sit for this portrait, destined to hang next to that of Cuvier in his gallery of eminent men in his own house. It is similar to another portrait by the same artist in the National Portrait Gallery, in which Owen holds an earlier anatomical triumph, the pearly nautilus shell.[48] But the most impressive of all the images of Owen is the 1896 black bronze statue by Thomas Brock, RA (1847–1922), located on the first landing above the main hall in The Natural History Museum. The gowned Professor stands on a plinth of Numidian marble and holds an unidentified moa bone in his left hand. At a comfortable distance behind Owen, and well out of sight, sit his two old adversaries, Charles Darwin and Thomas Huxley. The first of these was realised in white marble, by Sir Joseph Edgar Boehm (1834–90), and was presented by the Darwin Committee, of which Huxley was the chairman. But there was professional rivalry right to the end, for its unveiling had to wait until 1885, after Owen's retirement.[49] That of Huxley was sculpted in Carrara marble by Edward Onslow Ford (1852–1901) and unveiled in 1900.[50]

From his elevated position, Owen commands a grand sweep of the Main Hall. Directly in front of him is the tail-end of the plant-eating Diplodocus, a member of the family to which he gave the name 'dinosaur'. In an alcove off to one side of this giant creature is another, the moa, represented by the complete 2.68 metre tall skeleton of *Dinornis maximus*. According to the label, these are the bones of a 'feathered, flightless giant' which lived some 1000 years ago, and the last sighting of such a bird was 'about 1850'.

The very existence of The Natural History Museum is testimony to the energies and ambitions of Richard Owen. Deep within its Palaeontology Department's storage section lies collection item No. 44639, secure in its own customised box and cushioned on a bed of foam plastic beneath a clear acrylic lid.

Two accompanying paper labels record that this hollow section of the shaft of a left femur is the remains of *Dinornis novaezealandiae* Owen. One small handwritten note, 'Type of the Genus Dinornis', marks its unique place in history, as the first piece of a moa to be recorded by science. It is not at all difficult to see why this worn and discoloured scrap of bone was once dismissed as an 'unpromising fragment'.

# REFERENCES

## Introduction

1. Sir George Grey, letter to Richard Owen, 1849, quoted by E. Stewart, *New Zealand Herald*, 21 October 1977.
2. Atholl Anderson, *Prodigious Birds: Moas and Moa-hunting in Prehistoric New Zealand*, Cambridge University Press, England, 1989, Dedication (vii).
3. *Otago Daily Times*, 17 February 2003.

## 1. An Unpromising Fragment

1. Richard Owen, 'Exhibition of a Bone of an Unknown Struthious Bird from New Zealand', *Proceedings of the Zoological Society of London*, 7: 169–71, 1840.
2. 'London, 1753', The British Museum News Release, December 2001/67. Peter Ackroyd, *London: The Biography*, Vintage, Random House, London, 2001, 233–4.
3. Ackroyd, 2001, 229–37.
4. Jessie Dobson, *A Guide to the Hunterian Museum* (Physiological Series), The Royal College of Surgeons of England, Lincoln's Inn Fields, London, W.C. 1, 1958, 5–8.
5. A. W. Beasley, *Home Away from Home*, Central Institute of Technology, Wellington, in association with Grantham House, Wellington, New Zealand, 2000, 60, note 23.
6. Sir Victor Negus, *History of the Trustees of the Hunterian Collection*, E. & S. Livingstone, Edinburgh & London, 1966, 1–28.
7. *Illustrated London News*, October 4, 1845, 210.
8. *Synopsis of the Contents of the Museum of the Royal College of Surgeons*, R. & J. E. Taylor, Red Lion Court, Fleet Street, London, 1845. Ackroyd, 2001, 420.

# REFERENCES

9. *Synopsis of the Contents of the Museum of the Royal College of Surgeons*, R. & J. E. Taylor, Red Lion Court, Fleet Street, London, 1845.
10. Natural History Museum Archives, Owen Correspondence, vol. 22, no. 444(a).
11. Natural History Museum Archives, Owen Correspondence, vol. 22, no. 444(c), copy (by W. Clift) of original letter from J. W. Harris to Dr John Rule.
12. Sidney Lee (ed.), *Dictionary of National Biography*, Smith, Elder & Co., London, 1895, vol. xlii, 443.
13. Richard Owen, *Memoirs on the Extinct Wingless Birds of New Zealand; with an Appendix on those of England, Australia, Newfoundland, Mauritius, and Rodriguez*, John van Voorst, London, 1879, Preface (iii).
14. Owen, 1879, Preface (iii).
15. Owen, 'Notice of a Fragment of the Femur of a Gigantic Bird of New Zealand', *Transactions of the Zoological Society of London*, 3:29–32, 1842.
16. T. Lindsay Buick, *The Mystery of the Moa*, Thomas Avery & Sons Ltd, New Plymouth, New Zealand, 1931, 68. Owen also referred to Rule's bone as an 'unpromising specimen' in 1879, Preface (iii).
17. Professor Lord Zuckerman (ed.), *The Zoological Society of London: 1826–1976 and Beyond*, Academic Press, London, 1976, 1–5.
18. John Bastin, 'The First Prospectus of the Zoological Society of London: New Light on the Society's Origins', in *Journal of the Society for the Bibliography of Natural History*, vol. 5, 1968–71, Zoological Society of London.
19. Owen, 'Exhibition of a Bone of an Unknown Struthious Bird from New Zealand', *Proceedings of the Zoological Society of London*, 7:169–71, 1840.
20. Natural History Museum Archives, Owen Correspondence 90, no. 7.
21. Natural History Museum Archives, Owen Correspondence 90, vol. 2, no. 6.
22. Natural History Museum Archives, Owen Correspondence 90, no. 7.
23. Owen, 'Notice of a Fragment of the Femur of a Gigantic Bird of New Zealand', *Transactions of the Zoological Society of London*, 3:29–32, 1842.
24. Owen, 1879, Preface (v).

## 2. The Flax Factor

1. Letter from John Harris to John Rule, Sydney, 28 February 1837, Natural History Museum Archives, OC vol. 22, no. 444(c), copy by W. Clift.
2. Quoted in Anne Salmond, *Between Worlds*, Viking, Penguin Books, Auckland, 1997, 154.
3. Joseph Angus Mackay, *Historic Poverty Bay and the East Coast, N.I., N.Z.*, J. G. Mackay, Gisborne, New Zealand, 1949, 80.
4. Robert McNab (ed.), *Historical Records of New Zealand*,

Wellington, 1908, 36–7, 54–5, 195.

5. James Belich, *Making Peoples*, Penguin Books, Auckland, 1996, 128.

6. State Records of New South Wales, Colonial Secretary Index, 1788–1825; McNab, 1908, 323–7.

7. Robert McNab, *Murihiku and the Southern Islands*, William Smith, Invercargill, 1907, 145.

8. McNab, 1907, 199–210; McNab, 1908, 410–13, 457–74.

9. McNab, 1907, 199–210.

10. Lazarus Morris Goldman, *The History of the Jews in New Zealand*, A. H. & A. W. Reed, Wellington, 1958, 27–32.

11. Mackay, 1949, 99.

12. Thomas Lambert, *The Story of Old Wairoa*, Coulls Somerville Wilkie, Dunedin, 1925, 352.

13. Philip Whyte, 'Harris, John Williams', *Dictionary of New Zealand Biography*, Saturday, 1 December 2001.

14. Mackay, 1949, 101.

15. Mackay, 1949, 124.

16. Joel Polack, *New Zealand: Being a Narrative of Travels and Adventures during a Residence in that Country between the Years 1831 and 1837*, Richard Bentley, London, 1838, vol. 1, 257.

17. Mackay, 1949, 94.

18. Mackay, 1949, 145.

19. Natural History Museum Archives, Owen Correspondence, vol. 22, no. 444(c).

20. Richard Owen, 'Notice of a Fragment of the Femur of a Gigantic Bird of New Zealand',

*Transactions of the Zoological Society of London*, 3:29–32, 1842.

21. Mackay, 1949, 94.

22. Philip Whyte, 2001.

23. John Rule, 'New Zealand', *Polytechnic Journal*, 1843, no. 7, 8.

24. Natural History Museum Archives, Df105/11.

25. Professor Lord Zuckerman, *The Zoological Society of London: 1826–1976 and Beyond*, Academic Press, London, 1976, 87.

26. John Rule, 1843, 4.

## 3. Trade and Exchange

1. J. S. Polack, *New Zealand: Being a Narrative of Travels and Adventures During a Residence in That Country Between the Years 1831 and 1837*, Richard Bentley, London, 1838, Preface (iv).

2. Lazarus Morris Goldman, *The History of the Jews in New Zealand*, A. H. & A. W. Reed, Wellington, 1958, 33; Jocelyn Chisholm, 'Polack, Joel Samuel 1807–1882', *Dictionary of New Zealand Biography*, Saturday, 1 December 2001.

3. *Russell – More than a Place*, Russell Museum, 1997, 14.

4. J. S. Polack, *Manners and Customs of the New Zealanders: With Notes Corroborative of their Habits, Usages etc. and Remarks to Intending Emigrants*, James Madden & Co., London, 1840, vol. 2, 279.

5. A. H. McLintock (ed.), *An Encyclopaedia of New Zealand*, R. E. Owen, Government Printer, Wellington, 1966, vol. 1, 500.

6. Polack, 1838, Preface (iii).

7. McLintock (ed.), vol. 2, 790.

# REFERENCES

8. Polack, 1838, vol. 2, 329.
9. Polack, 1838, vol. 2, 274.
10. Polack, 1838, vol. 2, 117–8.
11. Polack, 1838, vol. 1, 346.
12. Polack, 1838, vol. 1, 345.
13. Polack, 1838, vol. 1, 303.
14. Polack, 1838, vol. 1, 307.
15. Polack, 1838, vol. 1, 307–8.
16. Polack, 1838, vol. 1, 308.
17. Polack, 1838, vol. 1, 302.
18. Polack, 1838, vol. 1, 307.
19. Polack, 1838, vol. 1, Preface (iii–v).
20. Polack, 1838, vol. 1, 347–8.
21. Polack, 1838, vol. 2, 146.
22. David Mackay, 'Colenso, William 1811–1899', *Dictionary of New Zealand Biography*, Saturday, 1 December 2001.
23. Letter, 22 August 1894, Mantell Collection, ATL; in Buick, 1936, 45.
24. Polack, 1838, vol. 2, 147.
25. Polack, 1838, vol. 2, 120.
26. Polack, 1838, vol. 2, 375–6; 1840, vol. 2, 143–4.
27. Polack, 1838, Appendix, note 11.
28. Goldman, 44–6.
29. Minutes of Evidence before Select Committee on the State of the Islands of New Zealand. Report from the Select Committee of the House of Lords, Appointed to Inquire into the Present State of the Islands of New Zealand and the Expediency of Regulating the Settlement of British Subjects Therein; with the Minutes of Evidence Taken Before the Committee, and an Index Thereto. Ordered, by the House of Lords, to be Printed, 8 August 1838. Parliamentary Archives, House of Lords Records Office, 49–55.
30. ibid., 79–95.
31. ibid., 111.
32. Goldman, 44; Polack, 1840, vol. 1, Introduction.
33. *Kidd's London Directory and Amusement Guide: A Hand-Book Showing How to Enjoy London in its Various Amusements, Exhibitions, Curiosities & C.*, W. Kidd, Covent Garden, London, (not dated).
34. Chisholm, 2001.
35. Polack, 1840, vol. 1, Introduction (xxvii–xxviii).
36. Polack, 1840, vol. 2, 282.
37. Polack, 1840, vol. 2, 281.
38. Polack, 1840, vol. 1, Introduction (xxv–xxvi).
39. William Frederick Poole, *Poole's Index to Periodical Literature*, Peter Smith, Gloucester, Mass., 1963.
40. *Eclectic Review*, 70:31 and 90: 414; *Monthly Review*, 147:161.
41. Walter E. Houghton (ed.), *The Wellesley Index to Victorian Periodicals 1824–1900*, University of Toronto Press & Routledge, 1988.
42. *Blackwood's Edinburgh Magazine*, vol. 43, March 1838, 371–84.
42. John Ward, *Information Relative to New Zealand: Compiled for the Use of Colonists*, John W. Parker, London, 1840, 2nd edition, Preface (vi).
43. H. Hill, 'The Moa – Legendary, Historical, and Geological: Why and When the Moa Disappeared', *New Zealand Institute Transactions and Proceedings*, vol. 46, 1913, 330.

44. Polack, 1838, vol. 2, opp. 120.

45. McLintock (ed.), 1966, vol. 1, 350.

46. Polack, 1838, vol. 1, 345; vol. 2, 389–90 and 1840, vol. 2, 267.

47. Chisholm, 2001.

48. Polack, 1840, vol. 1, Introduction.

49. McLintock (ed.), 1966, vol. 2, 789.

50. *New Zealand's Heritage*, Paul Hamlyn Ltd, Wellington, 1971, part 9, 243–6.

51. Joel Samuel Polack (worked 1823–30), 'Unknown Man', inscribed 1830, ivory 67 x 55. Bq. R. Garraway-Rice, 1262 Fiche 22/C3, P11-1933. Victoria and Albert Museum, London.

52. *Maori Bargaining with a Pakeha*, ink and wash drawing, 1845 or 1846. Version by Cyprian Bridge, unsigned, in the collection of the Alexander Turnbull Library, Wellington, A-079-017; version by John Williams, unsigned, private collection.

53. Hill, 1913, 346.

## 4. With God on Their Side

1. William Colenso, 'An Account of Some Enormous Fossil Bones, of an Unknown Species of the Class Aves, Lately Discovered in New Zealand', 1846, *Tasmanian Journal of Natural Science*, vol. 2, 81.

2. William Yate, *An Account of New Zealand and of the Church Missionary Society's Mission in the Northern Island*, 1835, London (facsimile edition by A. H. & A. W. Reed, Wellington, New Zealand, 1970), 58–61.

3. Judith Binney, 'Yate, William 1802–1877', *Dictionary of New Zealand Biography*, updated 30 September 2002.

4. Robert McNab (ed.), *Historical Records of New Zealand*, Wellington, 1908, 331–99.

5. McNab, 1908, 329, 331–99.

6. Philip Temple, *New Zealand Explorers: Great Journeys of Discovery*, Whitcoulls, Christchurch, 1985, 11.

7. J. L. Nicholas, *Narrative of a Voyage to New Zealand, Performed in the Years 1814 and 1815, in Company with the Rev. Samuel Marsden, Principal Chaplain of New South Wales*, James Black and Son, London, 1817, vol. 2, 255; H. Hill, The Moa – Legendary, Historical, and Geological: Why and When the Moa Disappeared', *New Zealand Institute Transactions and Proceedings*, vol. 46, 1913, 330.

8. Deborah Cadbury, *The Dinosaur Hunters*, Fourth Estate, London, 2001, 12.

9. *Quarterly Review*, vol. xl, 1852, 403.

10. Marie King, *A Most Noble Anchorage: A Story of Russell and the Bay of Islands*, Northland Historical Publications Society, 1992, 11–12.

11. King, 1992, 13–15.

12. Charles Darwin, *The Voyage of the 'Beagle'*, Everyman's Library, London, 1906, 401–11.

13. Darwin, 1906, 414.

14. Colenso, 1846, 81–2.

15. Richard Taylor, 'An Account of the First Discovery of Moa

# REFERENCES

Remains', *Transactions and Proceedings of the New Zealand Institute,* vol. v, 1872, 97.

16. Colenso, 1846, 85.
17. Colenso, *New Zealand Institute Transactions and Proceedings,* vol. 12, 1879, 106.
18. Taylor's Journal, MS.953, p. 156 (p. 111 typed transcript), Alexander Turnbull Library, Wellington, referred to in J. R. H. Andrews, *The Southern Ark: Zoological Discovery in New Zealand 1769–1900,* Century Hutchinson New Zealand Limited, Auckland, 1986, 127.
19. T. Lindsay Buick, *The Discovery of Dinornis: The Story of a Man, a Bone, and a Bird,* Thomas Avery & Sons, New Plymouth, 1936, 110: 'It is at least certain that in October, 1839, [Rule] was in London'.
20. Richard Taylor, *Journal,* 26 April 1839, Alexander Turnbull Library, Wellington, typescript II, 117, quoted in H. W. Orsman (ed.), *The Dictionary of New Zealand English,* Oxford University Press, New Zealand, 1997, 498.
21. Richard Taylor, 'An Account of the First Discovery of Moa Remains', *Transactions and Proceedings of the New Zealand Institute,* 1872, vol. v, 98.
22. Colenso, 1846, 85.
23. Colenso, 1846, 86.
24. Colenso, 1846, 87.
25. Colenso, 1846, 87–8.
26. A. C. Bagnall and G. C. Petersen, *Colenso,* A. H. & A. W. Reed, Wellington, 1948, 113.

27. Bagnall and Petersen, 1948, 119.
28. Bagnall and Petersen, 1948, 113–4.
29. Bagnall and Petersen, 1948, 115.
30. Bagnall and Petersen, 1948, 116.
31. Colenso, 'Memorandum of an Excursion, Made in the Northern Island of New Zealand, in the Summer of 1841–2; Intended as a contribution Towards the Ascertaining of the Natural Productions of the New Zealand Groupe: With Particular Reference to Their Botany', *Tasmanian Journal of Natural Science,* 1846, vol. 2, no. 8, 221–2.
32. Bagnall and Petersen, 1948, 128.
33. Bagnall and Petersen, 1948, 134.
34. Bagnall and Petersen, 1948, 135.
35. T. Lindsay Buick, *The Mystery of the Moa,* Thomas Avery & Sons, New Plymouth, 1931, 76.
36. Robin Woodward, *Cultivating Paradise: Aspects of Napier's Botanical History,* Hawke's Bay Museum, Napier, 2002, 38.
37. John Lillie, 'Introductory Paper', *Tasmanian Journal of Natural Science,* vol. 1, no. 1, 1842, 1.
38. Richard Taylor, The Bulrush Caterpillar', *Tasmanian Journal of Natural Science,* vol. 1, no. 4, 1842, 307–8.
39. Colenso, 1846, 81.
40. Colenso, 1846, 81–2.
41. Colenso, 1846, 89–90.
42. Colenso, 1846, 100.
43. Colenso, 1846, 101–2.
44. Colenso, 1846, 91–2.
45. Colenso, 'Description of Some New Ferns Lately Discovered in New Zealand', *Tasmanian Journal*

*of Natural Science,* 1842, vol. 1,
no. 5, 375.
46. H. Hill, 1913, 335–6.
47. Andrews, 1986, 129.

## 5. Dissension in the Ranks

1. Richard Taylor, 'An Account of
the First Discovery of Moa
Remains', *Transactions and
Proceedings of the New Zealand
Institute,* vol. v, 1872, 98.
2. J. D. Hooker, *Flora Novae
Zealandiae,* Introduction, 1852;
Robin Woodward, *Cultivating
Paradise: Aspects of Napier's
Botanical History,* Hawke's Bay
Museum, Napier, 2002, 36.
3. Hooker, 1852.
4. David Mackay, 'Colenso,
William 1811–1899', *Dictionary
of New Zealand Biography,*
Saturday, 1 December 2001.
5. William Colenso, 'Status Quo:
A Retrospect. – A Few More
Words by Way of Explanation
and Correction Concerning the
First Finding of the Bones of
the Moa in New Zealand; also
Strictures on the Quarterly
Reviewer's Severe and Unjust
Remarks on the Late Dr. G. A.
Mantell, F.R.S., &c., in
connection with the Same',
*New Zealand Institute Transactions
and Proceedings,* vol. 24, 1891,
473.
6. William Colenso, 'An Account
of Some Enormous Fossil Bones
of an Unknown species of the
Class *Aves,* Lately Discovered in
New Zealand', *The Annals and
Magazine of Natural History,* vol.
xiv, no. 89, August 1844, 81–96.
7. Colenso, 1891, 473.

8. Colenso, 1844, 81.
9. Colenso, 1844, 93 (Note A).
10. Colenso, 1844, 95.
11. Colenso, 1844, 90.
12. Colenso, 1844, 92.
13. Colenso, 1844, 93 (Note B).
14. Colenso, 1844, 96.
15. 'Note on the Moa Bones',
*Tasmanian Journal of Natural
Science,* vol. 2, no. 7, 106.
16. Richard Taylor, 'An Account of
the First Discovery of Moa
Remains', *Transactions and
Proceedings of the New Zealand
Institute,* vol. v, 1872, 98.
17. Taylor, 1872, 100.
18. Richard Taylor, 'An Account of
the First Discovery of Moa
Remains', *Transactions and
Proceedings of the New Zealand
Institute,* vol. v, 1872, 98.
19. Charles Darwin, *Journal of
Researches into Geology and
Natural History of the Various
Countries Visited by H.M.S. Beagle,
1832–1836,* 1839, quoted in
Nick Hazlewood, *Savage:
Survival, Revenge and the Theory of
Evolution,* Hodder and
Stoughton, London, 2001, 109.
20. Richard Taylor, *Te Ika a Maui, or
New Zealand and its Inhabitants,*
Wertheim and Macintosh,
London, 1855, 238.
21. Taylor, 1872, 97–8.
22. A. W. Beasley, *Home Away from
Home,* Central Institute of
Technology, Wellington, in
association with Grantham
House, New Zealand, 2000,
105–23.
23. Colenso, 'On the Moa', *New
Zealand Institute Transactions and
Proceedings,* vol. 12, 1879, 82.

24. Colenso, 1879, 103.
25. Abraham Rees ('with the assistance of Eminent Professional Gentlemen' and illustrated with numerous engravings 'by the most distinguished artists'), *The Cyclopaedia; or Universal Dictionary of Arts, Sciences, and Literature*, Longman, Hurst, Rees, Orme & Brown, Paternoster-Row, London, 1820.
26. Colenso, 1879, 103–4.
27. Woodward, 2002, 34.
28. Colenso, 1879, 106.
29. Colenso to Mantell, 22.8.1894, in Buick, 1936, 45.
30. Colenso, 1879, 101.
31. Colenso, 1891, 468.
32. *Quarterly Review*, vol. xc, 1852, 404–5.
33. Colenso, 1891, 470.
34. Gideon Mantell, 'On the Fossil Remains of Birds collected in New Zealand by Mr Walter Mantell', *Quarterly Journal of the Geological Society*, August, 1848, quoted in Colenso, 1891, 470–1.
35. Quoted in Colenso, 1891, 469.
36. Quoted in Colenso, 1891, 470.
37. Colenso, 1891, 473.
38. Richard Owen, *Memoirs on the Extinct Wingless Birds of New Zealand; with an Appendix on those of England, Australia, Newfoundland, Mauritius, and Rodriguez*, John van Voorst, London, 1879, Preface (v).
39. Colenso, 1891, 474.
40. Colenso, 1879, 104 (note).
41. Colenso, 1891, 476.
42. Owen, 1879, Preface (v).
43. Colenso, 1891, 475.
44. Owen, 'Memoir on the Genus Dinornis', *Transactions of the Zoological Society of London*, 3:3, 75–6, 1844.
45. Colenso, 1891, 476–7.
46. Woodward, 2002, 33.
47. A. H. McLintock (ed.), *An Encyclopaedia of New Zealand*, R. E. Owen, Government Printer, Wellington, 1966, vol. 1, 378–9.
48. *New Zealand Institute Transactions and Proceedings*, vol. 27, 1894, 688–9.
49. A. L. Rowse, *The Controversial Colensos*, Dyllansow Truran, Cornwall, 1989, 89.
50. A. G. Bagnall and G. C. Petersen, *William Colenso*, A. H. & A. W. Reed, Wellington, 1948, 468.
51. Oliver Stead (ed.), *150 Treasures*, Auckland War Memorial Museum & David Bateman, Auckland, 2001, 78.

## 6. Give the Man a Bone

1. *Quarterly Review*, vol. xc, 1852, 403.
2. Richard Owen, 'Notes on the Birth of the Giraffe at the Zoological Society's Gardens . . .' *Transactions of the Zoological Society of London*, vol. 3:21, 1849.
3. *Annals of Natural History*, vol. 2, no. 11, January 1839, 376–7.
4. 'Progress of Comparative Anatomy', *Quarterly Review*, vol. xc, 1852, 364.
5. Sidney Lee (ed.), *Dictionary of National Biography*, Smith, Elder & Co., London, 1895, 435–44.
6. A. W. Beasley, *Home Away from Home*, Central Institute of

Technology, Wellington, New Zealand in association with Grantham House, 2000, 79.

7. Sir Victor Negus, *History of the Trustees of the Hunterian Collection*, E. & S. Livingstone, Edinburgh & London, 1966, 18, 24, 118.

8. Ann Moyal, *Platypus*, Smithsonian Institution Press, Washington, D. C., 2001, 74–6.

9. Richard Owen, *Memoir on the Pearly Nautilus (with Illustrations of its External Form and Internal Structure, drawn up by Richard Owen)*, Royal College of Surgeons, London, 1832.

10. Lee (ed.), 1895, 436.

11. Jacob W. Gruber and John C. Thackray, *Richard Owen Commemoration: Three Studies*, Natural History Museum Publications, London, 1992, Preface (viii–ix); Negus, 1966, 32, 115.

12. J. R. H. Andrews, *The Southern Ark: Zoological Discovery in New Zealand 1769–1900*, Century Hutchinson New Zealand Ltd, Auckland, 1986, 67–70.

13. Sidney Lee (ed.), *Dictionary of National Biography*, Smith, Elder & Co., London, 1898, vol. liv.

14. Andrews, 1986, 70–3.

15. 'On the Anatomy of the *Apteryx Australis*, Shaw', *Transactions of the Zoological Society of London*, 1849, vol. 3:277–301.

16. Owen, 10th lecture, 29 May 1838, 'Digestive System of Birds'. Notes made by Wm. W. Cooper, later revised and corrected by Owen, Collection of Royal College of Surgeons, 42.C.26.

17. Owen, 1st lecture, 30 April 1839, 'Introduction to the Course'. Notes made by Wm. W. Cooper, later revised and corrected by Owen, Collection of Royal College of Surgeons, 42.C.27.

18. *Proceedings of the Zoological Society of London*, 1839, 7:63–5.

19. Leslie Stephen (ed.), *Dictionary of National Biography*, Smith, Elder & Co., 1888, vol. xiii, 317.

20. *A List of the Members of the Royal College of Surgeons in London*, various editions 1806–44, Collection of Hunterian Library, Royal College of Surgeons.

21. Richard Owen, *Memoirs on the Extinct Wingless Birds of New Zealand; with an Appendix on those of England, Australia, Newfoundland, Mauritius, and Rodriguez*, John van Voorst, London, 1879, Preface (iv).

22. Stephen Jay Gould, 'An Awful Terrible Dinosaurian Irony', in *The Lying Stones of Marrakesh: Penultimate Reflections in Natural History*, Vintage, London, 2001, 183–7.

23. Richard Owen, 'Memoir on the Genus Dinornis', *Transactions of the Zoological Society of London*, 3(3), 1844, Introduction (73).

24. Auckland Central Library, Grey Letters, vol. 30: GL-O10(3) & GL-O10(4).

25. Owen, 1844, Introduction (74).

26. Section of letter reproduced in Buick, *The Mystery of the Moa*, Thomas Avery & Sons Ltd, New Plymouth, 1931, 88–9.

# REFERENCES

27. ibid., 88–9.
28. Buick, 1931, 80.
29. *Transactions of the Zoological Society of London*, 1844, 3:243–75.
30. Buick, 1931, 321 (note 77).
31. Williams' letter to Buckland, quoted in Owen, 1844.
32. L. R. Brightwell, *The Zoo You Knew*, Basil Blackwell, Oxford, 1936, 50–1.
33. *Quarterly Review*, vol. xc, 1852, 402–3.
34. *Quarterly Review*, vol. xc, 1852, 403.
35. Negus, 1966, 118.
36. Unidentified publication, possibly a geological encyclopaedia, vol. xxv, 506–8, Natural History Museum Archives, Df105/11.
37. *Quarterly Review*, vol. xc, 1852, 402.
38. *Tasmanian Journal of Natural Science*, vol. 2, 107.
39. William Colenso, 'An Account of Some Enormous Fossil Bones, of an Unknown Species of the Class Aves, Lately Discovered in New Zealand', 1846, *Tasmanian Journal of Natural Science*, vol. 2, 101–2.
40. *Tasmanian Journal of Natural Science*, vol. 2, 1846, 106–7.
41. *Tasmanian Journal of Natural Science*, vol. 2, 1846, no. 6, 239–40: Reprinted from *Athenaeum*, no. 850, 138.
42. *Tasmanian Journal of Natural Science*, vol. 2, no. 10, 1845; Reprinted from *Athenaeum*, no. 882, 860.
43. *Annals of Natural History*, vol. v, no. xxvi, 147–52.

## 7. The Lone Fragment

1. Letter from Richard Bright to Richard Owen, 11 June 1870, Natural History Museum Archives, OC vol. 5, no. 59.
2. T. Lindsay Buick, *The Discovery of Dinornis: The Story of a Man, a Bone, and a Bird*, Thomas Avery & Sons Ltd, New Plymouth, 1936, 92–110.
3. Richard Owen, *Memoirs on the Extinct Wingless Birds of New Zealand; with an Appendix on those of England, Australia, Newfoundland, Mauritius, and Rodriguez*, John van Voorst, London, 1879, Preface (v).
4. Nicolaas A. Rupke, *Richard Owen: Victorian Naturalist*, Yale University Press, New Haven, 1994, 19.
5. *Geological Magazine*, no. cxii, October 1873, p. 478.
6. Owen, 1879, Preface (iv).
7. Letter from John Rule to Robert Bright, 16 February 1841, Natural History Museum Archives, Df105/11.
8. Owen, 'Exhibition of a Bone of an Unknown Struthious Bird from New Zealand', *Proceedings of the Zoological Society of London*, 7:169–171, 1840.
9. Letter from John Rule to Robert Bright, 27 March 1841, Natural History Museum Archives, Df105/11.
10. Letter from John Rule to Benjamin Bright, 31 March 1841, Natural History Museum Archives, Df105/11.
11. Owen, 1879, Preface (iii).
12. British Museum, Ethnology

Department Register
1757–1878.

13. J. Rule, 'New Zealand',
Polytechnic Journal, 1843, 7:1–13.

14. Owen, 'Notice of a Fragment of
the Femur of a Gigantic Bird of
New Zealand', Transactions of the
Zoological Society of London, 1842,
3:29–32.

15. Buick, 1936, 82–3.

16. E. Cecil Curwen (ed.), The
Journal of Gideon Mantell:
Surgeon and Geologist, Covering
the Years 1818–1852, Oxford
University Press, London,
1940, 225.

17. Buick, 1936, Preface (vi–vii).

18. J. E. Traue, 'Buick, Thomas
Lindsay', Dictionary of New
Zealand Biography, Saturday, 1
December 2001.

19. Buick, 1936, 50.

20. Buick, 1936, 51.

21. Transactions of the New Zealand
Institute, vol. iv, 1871, 379;
quoted in Buick, 1936, 53–4.

22. Buick, 1936, 67.

23. Buick, 1936, 58.

24. Buick, 1936, 55–61.

25. Buick, 1936, 67.

26. Buick, 1936, 88–92.

27. Sidney Lee (ed.), Dictionary of
National Biography, Smith, Elder
& Co., London, 1897, vol. xlix,
394–5.

28. Buick, 1936, 90.

29. Dictionary of National Biography,
1897, 394–5.

30. R. M. Kark and D. T. Moore,
'The Life, Work, and Geological
Collections of Richard Bright,
M.D. (1789–1858); with a Note
on the Collections of Other
Members of the Family', Archives

of Natural History, 10(1), 1981,
119–51.

31. Simon Winchester, The Map
that Changed the World, Penguin/
Viking, England, 2001.

32. F. W. Pledge, Crawley: Glimpses
into the Past of a Hampshire Parish,
'Privately published', England,
1907.

33. Letter from Richard Bright to
Richard Owen, 1869, Natural
History Museum Archives, OC,
vol. 5, no. 58.

34. Letter from Richard Bright to
Richard Owen, 1871, Natural
History Museum Archives, OC,
vol. 5, no. 57.

35. Letter from Benjamin Bright to
Richard Owen, 1872, Natural
History Museum Archives, OC,
vol. 5, no. 52–3.

36. ibid.

37. Letter from Benjamin Bright to
Richard Owen, 1872, Natural
History Museum Archives, OC,
vol. 5, no. 51.

38. R. M. Kark and D. T. Moore,
1981, 141.

39. Letter from Benjamin Bright to
Richard Owen, 1873, Natural
History Museum Archives, OC,
vol. 5, no. 54.

40. F. W. Pledge, 1907.

41. Geological Magazine, no. cxii,
October 1873, 478.

42. T. Lindsay Buick, The Mystery of
the Moa: New Zealand's Avian
Giant, Thomas Avery & Sons
Ltd. New Plymouth, 1931, 319,
note 65.

## 8. Bound for the Antipodes

1. John Ward, Information Relative to
New-Zealand Compiled for the Use

# REFERENCES

*of Colonists*, 2nd edition, John W. Parker, London, 1840, 1.

2. Ward, 1840, 2nd edition, 115.
3. Patricia Burns (ed. Henry Richardson), *Fatal Success: A History of the New Zealand Company*, Heinemann Reed, Auckland, 1989, 18–21.
4. Ward, 1840, 2nd edition, 109.
5. Ward, 1840, 2nd edition, 115–6, 126–7.
6. A. H. McLintock (ed.), *An Encyclopaedia of New Zealand*, R. E. Owen, Government Printer, Wellington, 1966, vol. 3, 179.
7. Denis McLean, 'Dieffenbach, Johann Karl Ernst 1811–1855', *Dictionary of New Zealand Biography*, updated Saturday, 1 December 2001.
8. Burns, 1989, 86.
9. McLintock (ed.), 1966, vol. 3, 249; Ward, 1840, 3rd edition, 169, 188.
10. Ward, 1840, 3rd edition, 182.
11. Charles Heaphy, *Mt. Egmont Viewed from the Southward*, c. 1840, watercolour, Alexander Turnbull Library, Wellington.
12. Ernest Dieffenbach, *Travels in New Zealand with Contributions to the Geography, Geology, Botany and Natural History of that Country*, John Murray, London, 1843, vol. 1, 135, 140–1, 155.
13. Dieffenbach, 1843, vol. 2, 195.
14. *Colonial Gazette*, 18 September, quoted in Burns, 1989, 110.
15. Ward, 1840, 2nd edition, 135.
16. Ward, 1840, 2nd edition, 131–2.
17. Ward, 1840, 2nd edition, 140–1.
18. Ward, 1840, 2nd edition, 141–2.
19. Burns, 1989, 126–9.
20. Ward, 1840, 2nd edition, 3.

21. Charles Hursthouse, *An Account of the Settlement of New Plymouth*, Smith, Elder & Co., London, 1849, 11.
22. Deborah Cadbury, *The Dinosaur Hunters*, Fourth Estate, London, 2001, 263.
23. E. Cecil Curwen (ed.), *The Journal of Gideon Mantell: Surgeon and Geologist, covering the years 1818–1852*, Oxford University Press, London, 1940, 201.
24. Cadbury, 2001, 264.
25. Buick, 1931, 324 (Note 92).
26. T. Lindsay Buick, *The Mystery of the Moa*, Thomas Avery & Sons Ltd, New Plymouth, 1931, 108.
27. Buick, 1931, 109.
28. Christopher McGowan, *The Dragon Seekers*, Little, Brown, London, 2002, 96.
29. Buick, 1931, 111.
30. J. R. H. Andrews, *The Southern Ark: Zoological Discovery in New Zealand 1769–1900*, Century Hutchinson New Zealand Limited, Auckland, 1986, 141–9.
31. McLintock (ed.), 1966, vol. 2, 405.
32. Curwen, 1940, 245.
33. Jacob W. Gruber, 'The Moa and the Professionalising of New Zealand Science', *The Turnbull Library Record*, Alexander Turnbull Library, Wellington, New Zealand, vol. 20, no. 2, October 1987, 77.
34. Cadbury, 2001, 324.
35. M. P. K. Sorrenson, 'Mantell, Walter Baldock Durrant 1820–1895', *Dictionary of New Zealand Biography*, Saturday, 1 December 2001.

36 *The Athenaeum*, no. 871, July 6, 1844, 630.

## 9. Fossils, Frogs and Grains of Sand

1. Quoted in Gideon Mantell, *Thoughts on a Pebble, or, a First Lesson in Geology*, Reeve, Benham, and Reeve, Strand, London, 1849, 8th edition.
2. Christopher McGowan, *The Dragon Seekers*, Little, Brown, London, 2002, 51–3, 93–4.
3. Sidney Lee (ed.), *Dictionary of National Biography*, Smith, Elder & Co., London, 1893, vol. xxxvi, 99–100.
4. Gideon Mantell, *Thoughts on a Pebble, or, a First Lesson in Geology*, Reeve, Benham, and Reeve, Strand, London, 1849, 8th edition, 8, 10, 75.
5. Lee (ed.), 1893, 99.
6. Gideon Mantell, *The Wonders of Geology*, Relfe and Fletcher, Cornhill, London, 1838, 3, 5, 100, 104.
7. Lee (ed.), 1893, 99.
8. *The Athenaeum*, no. 871, July 6, 1844, 630.
9. E. Cecil Curwen (ed.), *The Journal of Gideon Mantell: Surgeon and Geologist, Covering the Years 1818–1852*, Oxford University Press, London, 1940, 201.
10. Jacob W. Gruber, 'The Moa and the Professionalising of New Zealand Science', *The Turnbull Library Record*, Alexander Turnbull Library, Wellington, New Zealand, vol. 20, no. 2, October 1987, 72.
11. Curwen (ed.), 1940, 215.
12. Curwen (ed.), 1940, 219.
13. Gruber, 1987, 73.
14. Curwen (ed.), 1940, 219.
15. Curwen (ed.), 1940, 215.
16. Lee (ed.), 1893, 99.
17. Curwen (ed.), 1940, Introduction (viii).
18. *Annals of Natural History*, vol. 2, no. viii, 1848, 53–62.
19. Curwen (ed.), 1940, 219, 228.
20. Curwen (ed.), 1940, 229.
21. Curwen (ed.), 1940, 221.
22. Curwen (ed.), 1940, 221.
23. *Annals of Natural History*, vol. 2, no. viii, 1848, 51–2.
24. Curwen (ed.), 1940, 225.
25. Curwen (ed.), 1940, 225.
26. Curwen (ed.), 1940, 231.
27. Curwen (ed), 1940, 229.
28. Curwen (ed.), 1940, 240.
29. Curwen (ed.), 1940, 232.
30. Curwen (ed.), 1940, 248, 282, 290.
31. Curwen (ed), 1940, 232–3.
32. Curwen (ed.), 1940, 245–7.
33. Curwen (ed), 1940, 225.
34. Curwen (ed.), 1940, 232.
35. Curwen (ed.), 1940, 256.
36. Curwen (ed.), 1940, 245.
37. John Thackray and Bob Press, *The Natural History Museum: Nature's Treasurehouse*, Natural History Museum, London, 2001, 100, 136.
38. Curwen (ed.), 1940, 250.
39. Curwen (ed.), 1940, 255.
40. *Annals of Natural History*, vol. 7, 1851, 229.
41. J. R. H. Andrews, *The Southern Ark: Zoological Discovery in New Zealand 1769–1900*, Century Hutchinson New Zealand Limited, Auckland, 1986, 136.
42. Curwen (ed.), 1940, 260.
43. Curwen (ed.), 1940, 260–1.

44. Andrews, 1986, 146; Richard Holdaway and Trevor Worthy, 'Lost in Time', *New Zealand Geographic*, no. 12, 1991, 58–9.
45. Curwen (ed.), 1940, 267.
46. Curwen (ed.), 1940, 260–2.
47. Curwen (ed.), 1940, 268.
48. Andrews, 1986, 75.
49. Curwen (ed.), 1940, 278.
50. Curwen (ed.), 1940, 280.
51. Lee (ed.), 1893, 99.
52. Curwen (ed.), 1940, 292.
53. Deborah Cadbury, *The Dinosaur Hunters*, Fourth Estate, London, 2001, 293–6; Lynn Barber, *The Heyday of Natural History*, Jonathan Cape, London, 1980, 177.
54. Leslie Stephen (ed.), *Dictionary of National Biography*, Smith, Elder & Co., London, 1886, vol. vii, 206–8.
55. William Buckland, *Geology and Mineralogy Considered with Reference to Natural Theology*, William Pickering, London, 1836.
56. Buckland, 1836, 524–34, 595–7.
57. Stephen (ed.), 1886, 207.
58. Frank Buckland, *Curiosities of Natural History*, Herbert Strang's Library, London, n.d., Introduction (v).
59. Buckland, n.d, 7–9.
60. C. M. Gordon, *The Taylors and Bucklands: Early Auckland Pioneer Families*, published by C. M. Gordon, Auckland, New Zealand, 1969; Nancy M. Buckland, *The Bucklands (A Continuation of C. M. Gordon's Story and a Revision of the Buckland and Taylor Family Trees)*, published by Nancy M.

Buckland, 1975; Kenneth A. J. Buckland, *Caudebec – Buckland: Story of a Family 1065–1975*, published by Kenneth A. J. Buckland, 1977.

**10. In Search of Natural Knowledge**

1. F. W. Hutton, On the Moas of New Zealand, *Transactions of the New Zealand Institute*, 24, 1892, 93–171.
2. Peter Aughton, *Endeavour: The Story of Captain Cook's First Great Epic Voyage*, The Windrush Press, Gloucestershire, 1999, 112.
3. Aughton, 1999, 80–4.
4. Adrian Desmond and James Moore, *Darwin*, Penguin Books, England, 1991, 610.
5. William Colenso, 'On the Botany of the North Island of New Zealand', *New Zealand Institute Transactions and Proceedings*, vol. 1, part III, 1868.
6. J. R. H. Andrews, *The Southern Ark: Zoological Discovery in New Zealand 1769–1900*, Century Hutchinson New Zealand Limited, Auckland, 1986, 70.
7. F. Bruce Sampson, *Early New Zealand Botanical Art*, Reed Methuen, 1985, 47–8.
8. C. A. Fleming, *Science, Settlers, and Scholars: The Centennial History of the Royal Society of New Zealand*, The Royal Society of New Zealand, Wellington, 1987, 8.
9. Fleming, 1987, 7.
10. Walter Mantell, 'On Moa Beds', *Transactions and Proceedings of the New Zealand Institute*, vol. v, 1872, 94.

11. Fleming, 1987, 8.
12. Jacob W. Gruber, 'The Moa and the Professionalising of New Zealand Science', *The Turnbull Library Record*, Alexander Turnbull Library, Wellington, New Zealand, vol. 20, no. 2, October 1987, 64, 67–8.
13. Letter from Richard Owen to George Grey, 8 May 1839, Auckland Central Library, George Grey Correspondence, GL-O10(1), vol. 30, 61–4.
14. Letter from Richard Owen to George Grey, 7 November 1845, Auckland Central Library, George Grey Correspondence, GL-O10(3), vol. 30, 69–71.
15. Letter from Richard Owen to George Grey, 19 January 1847, Auckland Central Library, George Grey Correspondence, GL-O10(4), vol. 30, 72–4.
16. Letter from Richard Owen to George Grey, 26 January 1848, Auckland Central Library, George Grey Correspondence, GL-O10(5), vol. 30, 75–7.
17. Letter from Richard Owen to George Grey, 8 June 1850, Auckland Central Library, George Grey Correspondence, GL-O10(6), vol. 30, 78–81.
18. Letter from Richard Owen to George Grey, 11 November 1850, Auckland Central Library, George Grey Correspondence, GL-O10(7), vol. 30, 82–3.
19. E. Stewart, *New Zealand Herald*, 21 October 1977.
20. see 17.
21. Letter from Charles Darwin to George Grey, 13 November 1847, Auckland Central Library, George Grey Correspondence, GL-D8(2).
22. see 18.
23. Letter from Thomas Ralph to Richard Owen, 5 April 1852, Natural History Museum Archives, OC vol. 22, no. 44.
24. Ross Galbraith, *Walter Buller: The Reluctant Conservationist*, GP Books, Wellington, 1989, 33.
25. British Museum Ethnology Department, Register 1757–1878, 1854.
26. C. A. Fleming, 'Hochstetter, Christian Gottlieb Ferdinand von 1829–1884', *Dictionary of New Zealand Biography*, updated Saturday, 1 December 2001.
27. Ferdinand von Hochstetter, *New Zealand: Its Physical Geography, Geology and Natural History with Special Reference to Results of Government Expeditions to the Provinces of Auckland and Nelson*, J. G. Cotta, Stuttgart, 1867 (German edition, 1863), 184.
28. Hochstetter, 1867, 184.
29. H. F. von Haast, *The Life and Times of Sir Julius von Haast*, H. F. von Haast (publisher), Wellington, 1948, 32, 255.
30. Haast, 1948, 28.
31. Haast, 1948, 220.
32. Haast, 1948, 224.
33. T. Lindsay Buick, *The Mystery of the Moa*, Thomas Avery & Sons Ltd, New Plymouth, 1931, 130–1.
34. Haast, 1948, 602.
35. Fleming, 2001.
36. Hochstetter, 1867, Preface (v).
37. Hochstetter, 1867, opposite 176.

# REFERENCES

38. Atholl Anderson, *Prodigious Birds: Moas and Moa-hunting in Prehistoric New Zealand,* Cambridge University Press, England, 1989, 80–1.
39. Hochstetter, 1867, 191.
40. Hochstetter, 1867, 196.
41. Hochstetter, 1867, 178.
42. Hochstetter, 1867, 182.
43. Haast, 1948, 252–3.
44. Fleming, 1987, 17.
45. Hector, *New Zealand Institute Transactions and Proceedings,* vol. 1, 1868, Preface (i–iii).
46. Haast, 'Address – Delivered to the Philosophical Institute of Canterbury on 8 October 1869, Being the Anniversary of Captain Cook's 1st Landing in New Zealand', *New Zealand Institute Transactions and Proceedings,* 1869, 421–4.
47. Haast, 1948, 674–5.
48. Haast, 1948, 918–22.
49. Anderson, 1989, 102–6.
50. Haast, 1948, 666.
51. F. W. Hutton, 'On the Moas of New Zealand', *Transactions of the New Zealand Institute,* vol. 24, 1892, 97–8.
52. Hutton, 1892, 149.
53. Alfred Russel Wallace, *The Wonderful Century: The Age of New Ideas in Science and Invention,* Swan Sonnenschein & Co. Ltd, London, 1903, 385–6.
54. Haast, 1948, 666.
55. H. N. Parton, 'Hutton, Frederick Wollaston 1836–1905', *Dictionary of New Zealand Biography,* updated 11 December 2002.
56. W. P. Morrell, *The Anglican Church in New Zealand: A History,* Anglican Church of the Province of New Zealand, Dunedin, 1873, 87.
57. H. N. Parton, 'Bickerton, Alexander William 1842–1929', *Dictionary of New Zealand Biography,* updated 11 December 2002.
58. *Transactions of the New Zealand Institute,* vol. 27, 1894, 544–5.
59. *Transactions of the New Zealand Institute,* vol. 27, 1894.
60. Haast, 1948, 975–1022.
61. Haast, 1948, 225–6.
62. Richard Taylor, *Te Ika a Maui, or New Zealand and its Inhabitants,* William Macintosh, London & H. Ireson Jones, Wanganui, New Zealand, 1870, 603.

## 11. Extinguished Features

1. Thomas William Porter, 'Ardgowan', in *Legends of the Maori and Personal Reminiscences of the East Coast of New Zealand,* L. M. Isitt Ltd, Christchurch, 1925, 92.
2. John Ward, *Information Relative to New Zealand, Compiled for the Use of Colonists,* John W. Parker, London, 1840, 2nd edition, Preface (vi).
3. Ward, 1840, 52.
4. Richard Holdaway and Trevor Worthy, 'Lost in Time', *New Zealand Geographic,* no. 12, October–December 1991, 65.
5. *New Zealand Herald,* 1 November 2002.
6. *New Zealand Herald,* 9–10 November 2002 and 9 January 2003.
7. H. F. von Haast, *The Life and Times of Sir Julius von Haast,* H. F.

von Haast, Wellington, 1948, 810–1.

8. Te Rangi Hiroa Sir Peter Buck, *The Coming of the Maori*, Maori Purposes Fund Board, Wellington, 1958, 358–9.

9. Johannes C. Anderson, *Myths and Legends of the Polynesians*, Charles E. Tuttle Company, Rutland, Vermont, 1969, 139.

10. Anderson, 1969, 140–5.

11. Graeme Stevens, *Prehistoric New Zealand*, Heinemann Reed, Auckland, 1988, 58–9.

12. Haast, 1948, 579–84.

13. Julius Haast, 'Notes on a Collection of Saurian Remains from the Waipara River Made by Mr. J. Cockburn Hood and Shipped by Him to England on the Ship Mataoka for reference to Professor Owen', *New Zealand Institute Transactions*, vol. 2, 1869, 186.

14. James Hector, *New Zealand Institute Transactions*, vol. 6, 1873, 334.

15. Elsdon Best, *Forest Lore of the Maori*, E. C. Keating, Government Printer, Wellington, 1977, 183.

16. Best, 1977, 182.

17. A. W. Reed, *Myths and Legends of Maoriland*, A. H. & A. W. Reed, Wellington, 1961, 189–90.

18. Reed, 1961, 190–3.

19. A. W. Reed, *Treasury of Maori Folklore*, A. H. & A. W. Reed, Wellington, 1963, 288.

20. Reed, 1963, 289–90.

21. Porter, 1925, 33–6.

22. William Colenso, 'On the Maori Races of New Zealand', *New Zealand Institute Transactions and Proceedings*, 1869, vol. 1, 339–423.

23. William Colenso, 'On the Moa', Part II, 'What I have Gleaned Since', *New Zealand Institute Transactions and Proceedings*, vol. 12, 1879, 80–108.

24. Colenso, 1879, 80–1.

25. Douglas G. Sutton (ed.), *The Origins of the First New Zealanders*, Auckland University Press, 1994, 152–4; Ernest Dieffenbach, *Travels in New Zealand with Contributions to the Geography, Geology, Botany and Natural History of that Country*, John Murray, London, vol. 1, 1843, 368.

26. Colenso, 'An Account of Some Enormous Fossil Bones of an Unknown Species of the Class Aves, Lately Discovered in New Zealand', *The Annals and Magazine of Natural History*, no. 89, August 1844, 91.

27. Colenso, 'Contributions Towards a Better Knowledge of the Maori Race', *New Zealand Institute Transactions and Proceedings*, 1879, vol. 12, 110.

28. Colenso, 1879, 85–6.

29. Colenso, 1879, 89.

30. Colenso, 1879, 90–1.

31. Colenso, 1869, 339–423.

32. Richard Taylor, *Te Ika a Maui, or, New Zealand and its Inhabitants*, William Macintosh, London & H. Ireson Jones, Wanganui, New Zealand, 1870, 299.

33. James Stack, 'Notes on the Word "Moa" in the Poetry of the New Zealanders', *New Zealand Institute Transactions and*

# REFERENCES

*Proceedings*, vol. 7, 1874, Appendix (xxvii–xxix).

34. Elsdon Best, *The Maori*, The Polynesian Society, Wellington, 1941, vol. 1, xi; vol. 2, 487.
35. William Gilbert Mair, 'On the Disappearance of the Moa', *Transactions of the New Zealand Institute*, vol. 22, 1889, 70–5.
36. F. W. Hutton, 'On the Moas of New Zealand', *Transactions of the New Zealand Institute*, vol. 24, 1892, 160, 169.
37. Hutton, 1892, 162. Colenso, 1879, 94–7. Atholl Anderson, *Prodigious Birds: Moas and Moa-Hunting in Prehistoric New Zealand*, Cambridge University Press, England, 1989, 99.
38. Richard Holdaway, 'Terror of the Forests', *New Zealand Geographic*, 1989, no. 10, 56–65.
39. Richard Holdaway and Trevor Worthy, 'Lost in Time', *New Zealand Geographic*, no. 12, 1991, 51–68.
40. Sutton (ed.), 1994, 144, 150, 189, 251.
41. James Belich, *Making Peoples*, Penguin Books, Auckland, 1996, 51.
42. Ward, 1840, 33.
43. Holdaway and Worthy, 1991, 66.
44. Belich, 1996, 44.
45. Tim Flannery, *The Future Eaters*, Reed New Holland, Sydney, 1994, 180–2.
46. Flannery, 1994, 255–6.
47. Peter D. Ward, *Rivers in Time: The Search for Clues to Earth's Mass Extinctions*, Columbia University Press, New York, 2000, 251–2, 259.
48. Anton Gill and Alex West, *Extinct*, Pan Macmillan Ltd, London, 2001, 9.
49. Flannery, 1994, 55.
50. Charles Darwin, *The Origin of Species*, Mentor edition, USA, 1958, 333.
51. Darwin, 1958, 187.
52. Rhys Richards, *Which Pakeha Ate the Last Moa?*, The Paremata Press, Paremata, NZ, 1986, 42.
53. *Fourth Reader*, The Imperial Readers, Southern Cross Series, Whitcombe & Tombs Ltd, New Zealand (n.d.), 198–205.
54. Richards, 1986, 18–27.

## 12. Creating Monsters

1. Wilhelmina Sherriff Elliot, 'Beautiful Zealandia!' (to be sung to the tune of 'Maryland! My Maryland!'), in *From Zealandia*, John M. Watkins, London, 1925.
2. John Rule, 'New Zealand', *Polytechnic Journal*, no. 7, 1843, 7.
3. Richard Owen, 'Memoir on the Genus Dinornis', *Transactions of the Zoological Society of London*, vol. 3, 1844.
4. Richard Taylor, 'The Geology of New Zealand', *New Zealand Magazine*, vol. 1, no. 2, 101–11.
5. *Annals of Natural History*, vol. 7, 1851, 76–8.
6. Richard Taylor, *Te Ika a Maui*, Wertheim & Macintosh, London, 1855, 238.
7. Taylor, *New Zealand Magazine*, November, 1862, 176–85.
8. Taylor, *Te Ika a Maui*, William Macintosh, London & H. Ireson Jones, Wanganui, New Zealand, 1870, 443.

9. Richard Taylor, 'An Account of the First Discovery of Moa Remains', *Transactions and Proceedings of the New Zealand Institute*, vol. 5, 1872, 98.

10. Atholl Anderson, 'Shortland, Edward 1812?–1893', *Dictionary of New Zealand Biography*, updated 30 September 2002.

11. Henry Tacey Kemp to Hocken, 8 July 1896, Thomas Morland, personal letters and documents, ms-0451, folder 5, Hocken Library, Dunedin.

12. Edward Shortland, *The Southern Districts of New Zealand: A Journal with Passing Notices of the Customs of the Aborigines*, Longman, Brown, Green, & Longmans, London, 1851, 152–3.

13. *Wises New Zealand Guide*, Wises Publications Limited, Auckland, 1987, 53.

14. Herbert Wendt, *Out of Noah's Ark* (first published as *Auf Noahs Spuren*, 1956), The Riverside Press, Cambridge, Mass., 1959, 287–9.

15. F. A. Pouchet, *The Universe: or the Infinitely Great and the Infinitely Little*, Blackie & Son, London, 1895, 175.

16. *Webster's International Dictionary of the English Language*, G. & C. Merriam Company, Springfield, Mass., 1901, 993.

17. Charles Goldie and Louis Steele, *The Arrival of the Maoris in New Zealand*, 1898, Auckland Art Gallery Toi o Tamaki.

18. Leonard Bell, *Colonial Constructs: European Images of the Maori 1840–1914*, Auckland University Press, Auckland,
1992, 155–6.

19. Richard Wolfe, 'A Terrible Bird: The Fall and Rise of the Moa in New Zealand Art', *Art New Zealand*, no. 82, 1997, 71. 'The Chase of the Moa', in Francis Brewer Lysnar, *The Dear Old Maori Land*, Brett Printing and Publishing Co. Ltd, Auckland, 1924, 100.

20. Arthur Henry Adams, 'Maoriland', *Maoriland, and Other Verses*, Sydney, 1899, 1.

21. Elliot, 'Beautiful Zealandia!', 1925.

22. Richard Wolfe, *Kiwi: More Than a Bird*, Random Century New Zealand Ltd, Auckland, 1991, 16–39.

23. *Fourth Reader*, The Imperial Readers, Southern Cross series, Whitcombe & Tombs Limited, New Zealand (series appeared from 1899), 204.

24. F. W. Hutton (ed.) and James Drummond (compiler), *Nature in New Zealand*, Whitcombe & Tombs Limited, New Zealand, 1902, 7–12.

25. *Annual Report of the Auckland Institute and Museum*, 1912–13, 10 and 1913–14, 8–9.

26. Roger Duff, *Pyramid Valley*, The Association of Friends of the Canterbury Museum, Christchurch, New Zealand, 1952, 15–16.

27. Roger Duff, *Moas and Moa-Hunters*, School Publications Branch, New Zealand Education Department, 1951, 139.

28. Duff, 1951, 150.

29. Duff, 1951, 140.

# REFERENCES

30. Duff, 1952, 20-2.
31. Duff, 1952, 6.
32. *Evening Star*, 16 June 1950, reported in J. Herries Beatty, *The Moa: When Did it Become Extinct?*, published by J. Herries Beatty, 1953, 45.
33. Wolfe, 1997, 72.
34. Duff, 1952, 22.
35. Atholl Anderson, *Prodigious Birds: Moas and Moa-hunting in Prehistoric New Zealand*, Cambridge University Press, England, 1989, 62.
36. Brian Gill, 'The Moa Exhibits', *Auckland War Memorial Museum News*, no. 45, March 1991, 2.
37. Richard Holdaway and Trevor Worthy, 'Lost in Time', *New Zealand Geographic*, no. 12, 1991, 55-7.
38. Graeme Stevens, *Prehistoric New Zealand*, Heinemann Reed, Auckland, 1988, 48-50.

### 13. Back to the Bone

1. 'Ballad of the Ichthyosaurus', *Punch*, 14 February 1885, 82.
2. Nicolaas A. Rupke, *Richard Owen: Victorian Naturalist*, Yale University Press, New Haven, 1994, 1-2, 353.
3. Rupke, 1994, 13-4.
4. *More Pleasant Mornings at the British Museum*, The Religious Tract Society, London, 1850, 1-2.
5. ibid., 42-3, 134-5.
6. *Quarterly Review*, John Murray, London, vol. xc, 1852, 397-9.
7. *Quarterly Review*, 1852, 398-400.
8. Sir Victor Negus, *History of the Trustees of the Hunterian Collection*,

E. & S. Livingstone Ltd., England, 1966, 31.
9. Negus, 1966, 36.
10. Negus, 1966, 42-3.
11. Lynn Barber, *The Heyday of Natural History 1820-1870*, Jonathan Cape, London, 1980, 175.
12. A. H. McLintock (ed.), *An Encyclopaedia of New Zealand*, R. E. Owen, Government Printer, Wellington, 1966, vol. 3, 334.
13. Prof. Lord Zuckerman, *The Zoological Society of London 1826-1976 and Beyond*, The Zoological Society of London, Academic Press, 1976, 59.
14. Sidney Lee (ed.), *Dictionary of National Biography*, Smith, Elder, & Co., London, 1895, vol. xlii, 437, 443.
15. Letter from Thomas S. Ralph, Secretary, New Zealand Society, to Richard Owen, 5 April 1852, Natural History Museum, Archives, OC vol. 22, no. 44.
16. Nicola McGirr, *Nature's Connections*, The Natural History Museum, London, 2000, 50.
17. John Thackray and Bob Press, *The Natural History Museum: Nature's Treasurehouse*, The Natural History Museum, London, 2001, 42-50.
18. Thackray and Press, 2001, 55-60.
19. Rupke, 1994, 34-5; Thackray and Press, 2001, 57, 94-5; McGirr, 2000, 51-2.
20. Lee (ed.), 1895, 438.
21. *Quarterly Review*, 1852, 363.
22. Lee (ed.), 1895, 444.
23. P. Chalmers Mitchell, *Centennial*

*History of the Zoological Society of London*, Zoological Society of London, 1929, 119.

24. Stephen Jay Gould, 'An Awful Terrible Dinosaurian Irony', in *The Lying Stones of Marrakesh: Penultimate Reflections in Natural History*, Vintage, London, 2001, 186.

25. Letter from Herbert Rix, Assistant Secretary, Royal Society, to Richard Owen, Natural History Museum Archives, OC vol. 22, no. 302.

26. Lee (ed.), 1895, 443.

27. Lee (ed.), 1895, 443.

28. Robert Anderson, *The Great Court and the British Museum*, The British Museum Press, London, 2000, 21–2.

29. Adrian Desmond and James Moore, *Darwin*, Penguin Books, London, 1992, 411, 431–2.

30. W. W. Rouse Ball, *Notes on the History of Trinity College, Cambridge*, Macmillan & Co., London, 1899, 161–2; Sidney Lee (ed.), *Dictionary of National Biography*, Smith, Elder, & Co., London, 1895, vol. xlv, 437.

31. Lee (ed.), 1895, 440.

32. Desmond and Moore, 1992, 478.

33. Desmond and Moore, 1992, 452, 505.

34. *Transactions of the Zoological Society of London*, vol. iv, 1850, 15.

35. Charles Darwin, *On the Origin of Species by Means of Natural Selection (or the Preservation of Favoured Races in the Struggle for Life)*, John Murray, London, 4th edition, 1866, xvii–xviii.

36. Charles Darwin, *The Origin of Species (by Means of Natural Selection or the Preservation of Favoured Races in the Struggle for Life)*, John Murray, London, 6th edition (Popular Impression), 1902, xxvi–xxvii.

37. Darwin, 1902, 485.

38. Darwin, 1902, 218.

39. McGirr, 2000, 59–63.

40. *The Geological Magazine*, no. cxiv, December 1873, 529.

41. Charles Darwin, *The Voyage of the 'Beagle'*, Everyman's Library, London, 1906, 464.

42. Letter from W. Robinson to Richard Owen, 1886, Natural History Museum Archives, OC vol. 22, no. 354.

43. Rupke, 1994, 2.

44. Negus, 1966, 69.

45. Negus, 1966, 60.

46. Adrian Desmond, *Huxley*, Penguin Books, London, 1998, 306.

47. Negus, 1966, 104.

48. John C. Thackray, *A Catalogue of Portraits, Paintings and Sculpture at The Natural History Museum, London*, Mansell Publishing, Strand, London, 1995, 32.

49. Desmond and Moore, 1992, 675.

50. Thackray and Press, 2001, 136–7.

# INDEX

# INDEX

# INDEX

huia, 123
Hunter, John, 13–14, 70, 81
Hunterian Museum, 14–16, 19, 62,
    70, 81–4, 93, 98, 127, 136, 139, 142,
    169, 187, 190, 201, 202–3, 214–15
Hutton, Frederick Wollaston, 145,
    155, 163–4, 166, 179–80; *Nature
    in New Zealand*, 192–3
Huxley, Thomas, 167, 208, 216
*Hylaeosaurus*, 88, 127-8

## I

ichthyosaurs, 171
*Iguanodon*, 88, 124, 127, 128, 135, 138,
    139–40
Imperial Geological Institution
    (Vienna), 158; Museum, 156
International Industrial Exhibition,
    Vienna, 1873, 161
Ipswich Museum, 95

## J

Jaeger, G., 158
Jenner, Edward, 70

## K

katipo spider, 154
kauri, 54; gum, 47, 76
Kawhia, 30
Kawiti, 47, 151
kiwi, 23, 39, 44, 45, 49–50, 52, 67,
    84–8, 94, 96, 117, 123, 130, 137,
    138–9, 148, 159, 201; New
    Zealand national identity and,
    192 – *see also* entries beginning
    *Apteryx*
Kororareka, 36–7, 39, 40, 46, 47, 71
kumara, legend of how it came to
    New Zealand, 173–4

## L

Linnean Society, 20, 130, 154
Literary and Scientific Institution
    (Nelson), 149
lizards, 170
Logan, Dr, 85
London Institution, 131
Lord, Simeon, 28
Lyell, Charles, Sir, 116, 124, 127, 131,
    136, 164, 167; *Principles in Geology*,
    99

## M

Magniani, Mr, 158
Mahia Peninsula, 174
Mair, William Gilbert, 178–9
Mantell, Ellen, 121–2
Mantell, Gideon, 126–140; claims
    Colenso discovered struthious
    nature of moa, 73–5; death, 123,
    139; discovery of dinosaurs, 88,
    89, 127–8, 135, 138; *Fossils of the
    South Downs*, 128; health, 133, 134,
    139; John Rule visits, 105; pioneer
    in geology, 99; Richard Owen,
    relationship with, 73–5, 105,
    132–8, 139; Royal Medal awarded,
    135; talks and lectures, 131, 132–3,
    134; *Thoughts on a Pebble*, 128, 129;
    Walter Mantell sends specimens
    to, 120, 121, 122–4, 130–1, 137,
    151–2; *Wonders of Geology*, 129–30,
    138
Mantell, Mount, 144
Mantell, Walter, arrives in New
    Zealand, 120–21, 130; collects
    moa bones at Awamoa, 159, 180;
    collects moa bones at Waikouaiti,
    123; collects moa bones at
    Waingongoro, 68, 121–2, 130–1,

INDEX

Queen Charlotte Sound, 116
Queensland Museum, 198

## R

Ralph, Thomas Shearman, 153–4
Rangihoua, 51
ratites, 22, 193 – *see also* names of
    individual species, e.g. emu
*Rees' Cyclopaedia*, 72
*Regnosaurus Northamptoni*, 133
religion and science, 140–2, 208–9
*Resolution*, 27, 148
rhea, 23, 67, 72, 94, 96
Rix, Herbert, 206
Rongo-Kako, 174–5
Ross, James Clark, 64, 75
Rotorua, 118, 171, 179
Rowse, A.L., 78
Royal Astronomical Society, 20
Royal College of Surgeons, 14, 17, 19,
    20, 23, 34, 47, 70, 81, 82, 83–4, 98,
    101, 104, 105, 139, 150, 152, 214
Royal Geographic Society, 116
Royal Geological Society, 187
Royal Institution of London, 94
Royal Medal, 135
Royal Society, 20, 65, 132, 133, 138,
    146, 203; Gideon Mantell elected
    Fellow, 127; *Philosophical
    Transactions*, 127, 206; William
    Colenso first NZ Fellow, 77
Royal Society of London for
    Improving Natural Knowledge,
    20, 146
Royal Statistical Society, 20
Ruakapanga, 173
Rule, John, article for *Polytechnic
    Journal*, 102–4, 105; birth and
    early career, 107–8; brings moa
    bone fragment to Richard Owen,

13, 15–20, 26, 34–5, 56, 70, 87–8,
    98, 106, 118, 190; Dieffenbach's
    book acknowledges, 118;
    emigrates to Australia, 97–8;
    John Harris gives Maori artefacts
    to, 32–3, 101–2, 103; John Harris
    gives moa bone fragment to,
    32–3; Richard Owen doesn't
    acknowledge, 23, 95, 105, 106,
    111, 125, 133; sells moa bone
    fragment, 25, 98–102; Thomas
    Buick researches identity of,
    106–7; visits Gideon Mantell, 105
Rule, William Harris, 108
Russell, John, Lord, 47, 151
Rutherford, Ernest, 165

## S

Scharf, George, 25, 103, 135–6
science and religion, 140–2, 208–9
sealers, 29–30, 84, 123, 184, 189, 190
Selwyn, George Augustus, Bishop,
    40, 76, 89, 90
Shaw, George, 84–5
Shortland, Edward, 189
Skinner, H.D., 170
Smith, William, 108, 140
*Snapper*, 29
Solander, Daniel, 27, 145, 147
South America, 67, 201
South Island, 38, 151, 163, 180, 181,
    189, 190; name, Cookland, 161;
    name, Middle Island, 161; name,
    New Munster, 103
Stack, James William, 178, 179
Stanley, Lord, 85, 130
Steele, Louis, 191, 192
Stewart Island, 28; name, New
    Leinster, 103
*Struthioniformes* order, 22, 38, 62, 67, 72